THE CLIMATE CAPER

THE CLIMATE CAPER

FACTS AND FALLACIES OF GLOBAL WARMING

GARTH W. PALTRIDGE

WITH A FOREWORD BY
CHRISTOPHER MONCKTON

TAYLOR TRADE PUBLISHING
A CONNOR COURT BOOK
Lanham • New York • Boulder • Toronto • Plymouth, UK

ACKNOWLEDGMENT: This book is based on a number of the author's public talks and written articles over the past decade. He would like to thank *Quadrant* magazine in particular for raising no objection to the use of material he has published in that journal.

Published by Taylor Trade Publishing
An imprint of The Rowman & Littlefield Publishing Group, Inc.
4501 Forbes Boulevard, Suite 200, Lanham, Maryland 20706
http://www.rlpgtrade.com

Estover Road, Plymouth PL6 7PY, United Kingdom

Distributed by National Book Network

First published in 2009 by Connor Court Publishing PTY Ltd

British Library Cataloguing in Publication Information Available

Library of Congress Cataloging-in-Publication Data Available
978-1-58979-548-8 (pbk : alk. paper)
978-1-58979-549-5 (electronic)

∞™ The paper used in this publication meets the minimum requirements of American National Standard for Information Sciences—Permanence of Paper for Printed Library Materials, ANSI/NISO Z39.48-1992.

Printed in the United States of America

CONTENTS

FOREWORD

by The Viscount Monckton of Brenchley

The scholar-poet Callimachus, who catalogued the great Library of Alexandria a quarter of a millennium before Christ, said that a big book was a bad book. Nearly all books on "global warming" are big books, and nearly all of them are bad, being vexatiously and aggressively biased in one direction or the other.

Dr. Paltridge, by contrast, has written a book that would have delighted Callimachus - small, good, and refreshingly impartial, on a subject usually more laden with poisonous prejudice than perhaps any other in the history of human thought.

In the courteous, wryly elegant manner that was the hallmark of the great scientists of old, Paltridge gallops authoritatively, accessibly and always cheerfully through the physics, economics, sociology, philosophy and theology of the new religion of "global-warming" catastrophism that has so rapidly, baselessly and expensively mesmerized the international *classe politique*. This is very much a book for the general reader, who should not be in the least deterred by the very simple equation and couple of graphs in the physics chapter.

The author knows what he is talking about. He has been an eminent scientist for more than 40 years. He was a Chief Research Scientist with the CSIRO Division of Atmospheric Research, and was involved with the World Meteorological Organization's bureaucracy that set up the World Climate Program in the 1970s. He was working with the National Climate Program Office in the US at the time of the emergence of the UN's notorious climate panel, the IPCC, in the late 1980s.

For 12 years before his retirement in 2002, he directed the Institute for Antarctic Studies at the University of Tasmania, and was Chief Executive Officer of the Antarctic Co-operative Research Centre, studying the role of Antarctica and the Southern Ocean in climate change. He is also a Fellow of the Australian Academy of Science.

Though his book nowhere says so, he is still publishing learned papers of great brilliance in the peer-reviewed scientific literature. His most recent paper, published in 2009, questioned the IPCC's conclusion that the gentle "global warming" caused by CO_2 will be amplified into dangerous climate change by increases in the concentration of water vapour in the upper atmosphere. He showed that the satellites and balloon-borne radiosondes which measure humidity high in the atmosphere do not reliably confirm the positive water-vapour temperature feedback suggested by the theoretical models programmed into the Playstations and X-box 360's of the IPCC. Indeed, the radiosonde measurements in particular – insofar as they have any credence – suggest quite the opposite. They suggest that water-vapour temperature feedback over the last 3 or 4 decades has been negative and would, if continued into the future, greatly reduce the warming caused by CO_2.

The book is chiefly concerned with how and why climatologists, and scientists generally, have gone overboard in their support for the disaster theory attached to "global warming". His conclusion is blunt: "Even accepting for the sake of argument that some significant degree of global warming may be observed in the future, it is certainly not the consensus of the majority of scientists that the actual impact on humans will be significant – or indeed that it will be detrimental." It is hard to imagine a conclusion less congenial to the ruthless, *dirigiste,* scientific authoritarianism of the IPCC than that.

Dr. Paltridge does not question – any more than any competent scientist questions or has ever questioned – the fact that enriching the atmosphere with greenhouse gases will cause *some* warming.

He points out, rightly, that the high priests of the new religion tend to conflate the real scientific consensus to the effect that carbon dioxide causes *some* warming (in fact, an elementary and proven result in the physics of radiative transfer) with the imagined, and imaginary, "consensus" that the warming will be both significant and dangerous. In this conclusion he is supported by a recent peer-reviewed paper by Klaus-Martin Schulte, who read 539 papers selected randomly by the search-phrase "global climate change" and published from 2004-2007. Schulte found not one that offered any evidence that any human effect on the climate would be in any way catastrophic.

The conclusion, then, is fascinating enough. But the chief glory of Dr. Paltridge's book is his shrewdly-observed description – illustrated by frequent anecdotes – of the process by which science in general, and climate science in particular, has become so politicized that it is no longer recognizable as science at all and is, to that extent, no longer fit for its purpose. The "global warming" theory has all the worst attributes of a religion, with none of its saving graces. As Dr. Paltridge puts it:

> Most scientists simply cannot believe that their colleagues would deliberately oversell a scientific conclusion for the benefit of a political cause. Dishonesty of that nature would fly in the face of everything that the rather idealistic typical scientist has been taught about his profession. It follows from this belief that the activism of certain climate scientists cannot really be activism, but instead must be a dispassionate statement of truth by members of the profession who happen to know more than anybody else.

Gone is the academic freedom to open every door of enquiry, to follow every dream, to hold any opinion – however contentious – as long as that opinion springs from the tireless, careful, diligent, iterative process of hypothesis, investigation, measurement and testing, testing, testing that is the scientific method.

Gone is the determined scepticism that TH Huxley said was the highest duty of the improver of natural knowledge, replaced today by that blind faith that he called "the one unpardonable sin", and by the "sort of virtuous materialism that ... will noiselessly unwind the springs of action" that de Tocqueville, almost two centuries ago, foresaw might emerge as a potentially fatal consequence of democracy itself.

Science is not a democracy, and it cannot and does not proceed by consensus. Today's notion that it can and does is merely an instance of the Aristotelian logical fallacy of the *argumentum ad populum* – the head-count fallacy.

Gone is the right of free speech that was until recently one of the proudest achievements of democracy, enshrined in the First Amendment to the US Constitution. Now, anyone who dares to speak as Dr. Paltridge has done – even if he speaks with the unfailing politeness and charm of Dr. Paltridge, and with his unquestionable authority and impartiality – is branded a "climate change denier", who, in the words of a recent and unlamented British Foreign Secretary, ought to be "treated like an Islamic terrorist and denied all access to the news media".

Dr. Paltridge begins and ends his book with a quotation from the farewell speech of Dwight D. Eisenhower on stepping down from the Presidency of the United States in 1961. That speech is best known for its warning against the power of the "industrial-military complex". Dr. Paltridge, however, draws attention to the second and less well-known warning given by the outgoing President – that politicians might find themselves deferring to a "scientific-technological elite".

In the past, politicians with no scientific training would at least have had a Classical education. As part of the *studium generale,* the core curriculum that was taught to all undergraduates at all universities from the Middle Ages until World War II, every aspiring politician would have learned to recognize the Aristotelian fallacies a mile off.

"The IPCC/NAS/Royal Society is august and respected, and believes climate change will be disastrous": that is the *argumentum ad verecundiam,* the fallacy of reputation. Two words demonstrate that fallacy: Bernie Madoff.

"Hurricane Katrina did a lot of damage, so 'global warming' is real": that is the improper argument from the particular to the general that is the fallacy of converse accident.

"There has been 'global warming', and we do not know why, so it is prudent to assume that it is caused by humankind": that is the *argumentum ad ignorantiam,* the fallacy of arguing from ignorance.

"Polar bears are cuddly, therefore they are threatened by 'global warming"': that is the *argumentum ad misericordiam,* the fallacy of drawing improper conclusions because one is blinded by taking pity on some creature or another. The polar bears, by the way, are doing fine. There are five times as many of them as there were in the 1940s – hardly the profile of a species in imminent danger of extinction.

No doubt the publication of Dr. Paltridge's book will be greeted by numerous personal attacks on him by true-believers in the new religion. If so, like many before him who have publicly dared to question the questionable, he will be repeatedly subjected to the *argumentum ad hominem,* the logical fallacy of criticizing the person rather than the case he argues. That has been the fate of most "climate deniers".

Nevertheless, when the history of the bizarre intellectual aberration that is "global warming" comes to be written, once it is even clearer than it already is that the disasters, catastrophes, cataclysms and apocalypses that have been so luridly and so widely predicted have not and will not come to pass, Dr. Paltridge's little book will be regarded as one of the few, rare, precious beacons of enlightenment that prevented humanity from wandering through carelessness, ignorance and absent-mindedness into a new Dark Age.

Monckton of Brenchley
Carie, Rannoch, April 2009

INTRODUCTION

About ten years ago I was at a meeting organized by some of the movers and shakers of Australian science and technology. They had invited various experts on greenhouse global warming to talk about the most recent report at that time of the Intergovernmental Panel on Climate Change (the IPCC). For the uninitiated, the IPCC is a body of scientists and international bureaucrats which, among other things, produces a report every four or five years on the science of global warming. The reports are enormously influential. They form the platform on which is based much of the world's activism on the greenhouse issue.

The experts at the meeting discussed in particular a new proxy record of the world's temperature over the last thousand years. The record was given prominent treatment by the IPCC, and was perhaps the most talked about aspect of its report. It was based mainly on the analysis of tree rings at a number of sites in the northern hemisphere. It became famously known as the 'hockey stick'. This was because the rise in measured temperature over the last hundred years seemed, by comparison with the proxy record, to be much greater than any changes over the previous 900 years. The graph indeed looked to have the shape of a hockey stick lying down with its blade turned up at the end. Here (at last?) seemed to be good experimental evidence that the world is rapidly warming as a result of man's profligate use of fossil fuel.

So as not to bore the socks off everyone, it is probably sufficient at this stage to say that the hockey stick analysis had rather a lot of problems. One of them concerned the matching of the instrumental record of the last hundred years with the proxy record of the previous nine hundred. Changing an instrument in the middle of a series of measurements is always a suspicious procedure, and is infinitely more so when the timing of the change corresponds exactly with the timing of an apparent change in the quantity being measured.

Then again, it is extraordinarily difficult to detect the rise in global temperature over the last hundred years even though there have been continuous measurements during that time by many thousands of calibrated thermometers spread all around the world. Scientists and statisticians are still arguing about the matter. How much more questionable must be a nine-hundred year record based on measurements at only a very few places - much less than a dozen for a lot of the time. And tree rings are scarcely calibrated thermometers. While ring thickness may indeed be governed by temperature in a rough sort of way, it is also influenced by lots of other things such as rainfall, disease and local topography.

Anyway it seemed reasonable at the meeting to point out some of these problems, and to suggest that perhaps the IPCC had gone a little overboard in its reporting of the matter.

It was like stirring a hornet's nest. One after another the global warming experts rose to condemn me for questioning in public the conclusions of an IPCC report that had been compiled and endorsed as the consensus opinion of a large number of knowledgeable scientists. What right had I to make negative comments when I was not an expert in that particular aspect of climate science? If I wanted to question the science then the proper procedure was to write my thoughts in a formal scientific paper that could be subjected to peer review. And so on and so forth. Suffice it to say that the verbal spat was quite out of proportion to whatever was the crime that had been committed. Through it all the bemused audience of movers and shakers sat quietly on the sidelines. Lord knows what they thought about it all. The condemnation in fact continued for some days afterwards with a rash of fairly rude e-mails (of course with copies to the movers and shakers) demanding that I apologize for bringing disrepute to the IPCC process and to the scientific personnel associated with it.

All of which is fairly petty stuff. It would not normally be worth reporting except that it is a small example of how difficult it can be these days for the ordinary scientist to question the official beliefs of the apparatchiks of global warming. The IPCC was always going to be a lobby mechanism for a particular view of the climate change issue, so it is not too surprising that it has become more than a little messianic and tends to ignore contrary opinion. Certainly its behaviour argues a belief in the old adage that the end justifies the means. But its most remarkable achievement is that it has introduced a sort of religious supportive fervour into the behaviour of many of the scientists directly involved in its activity.

A colleague of mine put it rather well. The IPCC, he said, has developed a highly successful immune system. Its climate scientists have become the equivalent of white blood cells that rush in overwhelming numbers to repel infection by ideas and results which do not support the basic thesis that global warming is perhaps the greatest of the modern threats to mankind.

For the record before proceeding, it turns out that the hockey-stick reconstruction of past climate is indeed fairly close to being nonsense, and for a reason quite different from any of those mentioned above. A couple of Canadian statisticians found that there had been all sorts of errors and omissions in the purely mechanical process of information assembly. When the analysis was done properly, the data in fact suggested that the fifteenth century was much warmer than today – a conclusion that sits rather badly with fashionable doomsday scenarios of future climate change. Not, it should be emphasized, that the revised picture should be taken too seriously. The basic measurements still have a lot of problems.

The main point of telling here the story of the meeting is that the occasion was the first time that I began really to appreciate that the climate-change bandwagon was going seriously off the rails of what one might call 'normal' science. It was becoming a very obvious example of the pursuit of political correctness – a pursuit in which climate scientists themselves were not only major players but also major drivers as well. And while they are always very careful to make the point, at least in the presence of their peers, that it is not them but the press and the politicians who are over-blowing whatever is the global warming disaster story of the day, it is fairly obvious that they rarely make an effort to correct the public impressions being so actively created.

The whole business has hardened over the last decade or so into a semi-religious crusade in which climate scientists have developed an arrogance about their aims and activity which brooks no argument either with their interpretation of the science or with the way in which the science is used. To achieve their ends, they are drawing heavily on the capital of scientific reputation that has been so painfully assembled over hundreds of years. And it is they, not the politicians or the press, who have set up the various international mechanisms such as the IPCC that are designed to keep the whole issue at the forefront of public attention. It is they who are publicly scornful of sceptical scientists from outside their field. It is they who privately ostracize any sceptical colleagues from within their own ranks.

The question is why. In this book we examine the climate-change issue from various points of view and, without providing a definitive answer to the question, try at least to set out some of the conditions that have allowed the present dangerous state of affairs to come into being.

CHAPTER 1

OVERVIEW

"Yet, in holding scientific research and discovery in respect, as we should, we must also be alert to the equal and opposite danger that public policy could itself become the captive of a scientific-technological elite."

From President Eisenhower's farewell address to the nation
January 17, 1961

The greenhouse global-warming issue has run much faster and further than a lot of scientists expected two decades ago when the first report by the Intergovernmental Panel on Climate Change – the IPCC for short – was being written. Normally there is a cycle of public and scientific interest in environmental issues, and we guessed at the time that the concern with global warming was already past its peak. The point we all missed was that the global warming issue could be manipulated into the ultimate example of the politically correct. The need to do something about the problem plays to the agendas of virtually all branches of modern social activism.

Suffice it to say that the science behind the issue, and particularly the uncertainty of the science behind the issue, was irrelevant even before the so-called 'IPCC process' got off the ground.

The organizers of the IPCC set up a report-producing mechanism involving three separate international working groups. The first dealt with the science behind the actual predictions of climatic warming.

The second concerned the potential impact of that warming on human society and its welfare. The third was the only group that really mattered. It was designed right from the outset to examine and recommend options for international action to avert a climatic disaster.

The groups did their work in parallel. There was no pretence by the semi-political international negotiators of the third working group of waiting for outcome from the deliberations of the scientists within the other two groups. There was no need. They could develop their arguments quite independently because they knew, more-or-less, what the scientists would say. In fact all they really needed to know was that the other two working groups would *not* come out with a categorical statement to the effect that greenhouse warming is a load of nonsense. In this respect the third working group was betting on an absolute certainty. Even if the science pointed towards the load-of-nonsense theory – which it didn't and doesn't – no scientist worth his or her salt would make a categorical statement to that effect. After all, the basic difficulty with the climate-change issue is that it is a conglomeration of uncertainties.

And so the die was cast and the climate-change juggernaut began to roll in earnest. The central thesis was, and still is, that the burning of fossil fuel increases the concentration of carbon dioxide in the atmosphere. This increase will induce a warming of the world because it 'acts like an extra blanket' and reduces the loss of radiant heat to space. The thesis goes on to say that the change to a warmer climate will alter the future circumstances of man to some degree as yet unknown. Some people may benefit and others may be disadvantaged. We cannot yet determine whether the world as a whole will benefit or be disadvantaged by climate change, and we certainly cannot yet determine who might be the specific winners or losers in the altered environment. Therefore humans should quickly limit those of their activities that contribute to the problem. In practice this means that we should massively reduce our consumption of fossil fuel.

This is something of a tall order in a world where the economic welfare of all nations is governed almost entirely by access to cheap energy.

Over the years since those first moves by the IPCC, we have been subjected to an ever-increasing number of high-level international conferences and meetings designed specifically to raise the consciousness of the world to the potential for climatic disaster. We have seen enormous sums of money appearing out of nowhere to support the efforts of environmental lobby groups in their efforts to influence governments away from dependence on fossil fuels. We have seen a deliberately fostered change of public opinion from one in which there was a possibility of detrimental climate change for some people to one of absolute certainty that there will be climatic disaster for everyone.

The campaign has been remarkably successful, and the nations of the world are now in the middle of developing their expensive strategies for dealing with the matter. It is difficult now to question either the policies or the premises on which those strategies are based. Scientists are told quite bluntly that it is completely inappropriate for them to speak on matters of policy – unless of course they support it. Non-scientists are assumed to be technically ignorant and thereby incapable of speaking with authority on the issue. The machine, in other words, has it both ways and can run free.

Well, not entirely free. There is still a need to keep the issue before the public because nations are getting to that awkward stage where real upheavals of national economies have to be contemplated if more than lip service is to be paid to the activists' desire for international action. For this purpose the scientists are inveigled into producing more reports, into holding more conferences, into doing more research. The IPCC has become a permanent feature on the landscape, with its own bureaucracy spawning a continuing series of detailed reports.

The science discussed in these reports is interesting enough but has not really changed our basic knowledge (or our basic ignorance) of the problem. It may never do so, because there are many aspects of climate that are inherently unpredictable. In all probability, nothing much will change until the scientific community either provides categorical predictions as to who will be winners and who will be losers, or comes around to the extreme 'load-of-nonsense' theory.

Neither outcome is likely. The political point of the scientific activity is that it lends respectability to the social and political machinations. The scientists themselves are kept happy with money, and have in any event become extremely good players of the political game. Most of the good ones are in government research laboratories, and know exactly how their bread is buttered.

Let us accept for the sake of argument that the scientific consensus is for a change of climate in the direction of a global warming. The fact that 'consensus' is scarcely the way to run the railroad of science now seems unimportant to everyone except a few old-fashioned researchers who fear for the future of their profession. The most important question now about climate change is "will it be big enough to matter?" And the ultimate question is "if it matters, is it worth doing anything about?" The politically correct attitude to these questions is the same as before. Namely, if we don't know what will be the outcome, then we should take action to stop it.

The trouble is that political correctness is not often a good guide either to common sense or to reasonable ethics. The 'take action now' argument is only completely valid if the possible outcomes of change are known to be disastrous for everyone. Provided one accepts that change of itself is not inherently bad, it is far more likely that the distribution of good versus evil to come out of climate change would be roughly fifty-fifty. So that an active program of altering our present behaviour so as to reduce potential climate change is as likely to cause harm as is a program of masterly inaction. Particularly if the active program costs money - and real money at that. Today's dollars, and not those dollars discounted virtually to zero for the future.

It is in fact extremely difficult to sustain an argument for active limitation of greenhouse warming purely on economic grounds. If for no other reason, the standard rates of discount for the future make sure of that. Indeed, from the selfish viewpoint of the wealthier, geographically larger and economically more diverse nations, it is relatively easy to sustain an argument that such limitation would be economically stupid. For such nations the climate-susceptible industries such as agriculture account for a small and diminishing percentage of gross national product. It does not make selfish economic sense (we are being economically rational for the moment) for such countries to make major changes to the energy industries which underpin a large fraction of gross national product in order to limit an impact on industries that contribute a very much smaller fraction.

The real problem comes with defining a universally acceptable discount rate for those natural environmental things which we all appreciate but on which we cannot put a value in dollars. The concepts of economically sustainable development get in the way – in particular the concept that we should 'behave in a way which meets the needs of the present without compromising the ability of future generations to meet their needs'. Since it is difficult to predict what might be the needs of future generations as far as the natural environment is concerned, the problem inevitably boils down to defining a universally acceptable discount rate for the present state. At which point there is a further problem of logic. It is believed that no natural ecosystem (in Australia certainly) is in steady state. Ecosystems change naturally whatever we do. Since we have no particular expectation of what the ultimate steady state of an ecosystem might be, there is little point in getting upset because it might alter as a result of a change in climate. How then can we talk about a 'discount rate for the present state'?

Where does all this get us? Not very far except to a general feeling that, from the point of view of countries like Australia, whether or not there might be significant greenhouse warming is probably not all that important in the overall scheme of things. If one is selfish enough to accept that, then it is possible also to accept two propositions. The first is a restatement of the obvious. If we are to do anything about restricting greenhouse gas emissions, then it should be done only if there are other compelling and shorter-term reasons for taking whatever might be the proposed action. Playing around with 'clean coal technology' – a pseudonym for expensively burying in the ground the carbon emitted by coal-fired power stations – doesn't fit that philosophy. The second is simply a statement of hardnosed international politics. If we are to spend resources on restricting our greenhouse gas emissions, then what is the rest of the world going to do in return? 'Taking a lead' with an expensive program to reduce carbon emission in the virtuous hope that other countries will be shamed into doing the same is not an obviously sensible procedure. Shame does not figure largely in the determination of national agendas.

CHAPTER 2

SOME PHYSICS

There are good and straightforward scientific reasons to believe that the burning of fossil fuel and the consequent increase in atmospheric carbon dioxide will lead to an increase in the average temperature of the world above that which would otherwise be the case. Whether the increase will be large enough to be noticeable is still an unanswered question. And we don't really know what would otherwise be the case. There is a good and straightforward scientific reason (the second law of thermodynamics no less!!) to expect that the average temperature of areas the size of the broad climatic regions of Australia will rise more-or-less in concert with the global average. There is no such reason to expect that other regional-scale climate variables – rainfall in particular – will change in any particular way with the change in global temperature. Indeed it is still theoretically possible that long-term changes in regional and continent-wide rainfall are inherently unpredictable because of the turbulent nature of the atmosphere and the ocean.

There is a fair amount of reasonable science behind the global warming debate, but in general, and give or take a religion or two, never has quite so much rubbish been espoused by so many on so little evidence. One wonders why. We live in an age where common sense and tolerance are supposed to be the basis of our system of education, but there is very little of common sense and absolutely nothing of tolerance in the public argument about the climate change business. Perhaps it is that people simply have a basic need for fairy tales and doomsday stories.

Let us begin by putting the 'reasonable science' in perspective, and think a little about a plume of smoke rising from a cigarette into some sort of flue.

The stream of smoke is smooth enough for a start, but suddenly breaks into turbulent eddies whose behaviour is inherently unpredictable. We can make closely-spaced measurements all over the plume at some particular time and predict things for a little while thereafter, but very soon random fluctuations smaller than the distance between the measurements (they are called 'sub-grid-scale eddies' in the vernacular of the numerical modellers) grow in size and appear out of nowhere to swamp the eddies we thought we knew something about. While we can say that, on average, the overall column of smoke will continue to rise, we can make that rather limited statement only because the random fluctuations cannot grow to a size any bigger than the dimension of the flue.

Figure 2.1

Predicting the actual rate of rise is more difficult. Depending on the shape of the flue, it may require a bit of cheating with an experiment and the use of a 'tuneable parameter' in the forecasting process. A tuneable parameter is a piece of input information whose actual value is chosen on no basis other than to ensure that theoretical simulation matches observation.

The climate system is much like the smoke but is vastly more complicated. The atmosphere and the ocean are two interacting turbulent media with turbulent processes going on inside them, and there are all sorts and shapes of physical boundary that may or may not allow prediction of average conditions over areas less than the size of the earth. In principle at least we may be able to make a reasonable forecast of such things as the future global-average temperature and global-average rainfall by using a numerical model and a fair number of tuneable parameters obtained from observations of present conditions. Forecasting smaller-scale averages becomes more and more problematic as the scale decreases. As a first guess, one might be able to forecast averages over areas the size of ocean basins, but one cannot really expect to make skilful prediction for areas much smaller than that.

This hasn't stopped people trying. It's a fairly popular technique these days in order to get over this business of the growth of sub-grid-scale eddies to run 'ensembles' of climate models. The typical model throws a dense computational grid over its simulated atmosphere and ocean, and it is set running into the future with initial conditions taken from measured data at each of the grid points. The process is repeated for lots of similar trials with small changes in the initial conditions, and then the researcher looks for what are called robust results. That is, he looks for averages of some variable or another over some particular area or another for which changes in the initial conditions from one trial to another don't seem to make much difference. These particular results at least, it is imagined, are robust and should represent what will happen in the real world.

Well maybe. There are a few things to think about.

There is a tendency to believe that, quite apart from the specific case of robust forecasts, the simple 'ensemble average' of the runs is the most likely to represent the future of the real atmosphere and ocean. There may be some truth to this, but it relies on a basic assumption to the effect that changes in 'discretely-sampled initial conditions' (that is, initial conditions defined and projected forward only at grid points which, in the case of climate models, are some tens or hundreds of kilometres apart) have the same sort of consequence as small-scale fluctuations in the real and continuous fluids. Do they really have the same sort of effect on future climate as the flapping of a few butterfly wings over Canberra next Thursday? And in any event, if, as some would have it, butterfly wings can indeed introduce massive changes in future climate, surely they will produce a particular distribution of climate that may have nothing to do with the ensemble average of a lot of grid-point models.

In the search for robust results one has to bear in mind the fundamental difference between a variable like temperature and another like rainfall. According to the Le Chatelier Principle, which is a sort of limited version of the second law of thermodynamics that some of us heard about at school in chemistry classes, temperature is highly conservative. That is, any physical system seems to work very hard to smooth out its own temperature fluctuations. As a consequence there is a natural tendency for complex systems to yield reasonable averages of temperature. Put another way, if the average temperature of the world were to increase then it is a fair bet that the temperature of the individual regions would increase more-or-less in concert with the average. On the other hand, nobody has yet come across a sort of Le Chatelier Principle concerning rainfall. It is not surprising that forecasts of rainfall – the most significant climate variable as far as many people are concerned – are spectacularly unsuccessful in practice, and spectacularly changeable and non-robust from one model to another.

To the extent that there are at least some large-scale areas over which averages of climate may be robust and predictable in principle, it seems reasonable to guess that there ought to be some cunning physics which can be applied to the climate system on that sort of scale. This rather than pursue ever-more resolution and ever more detail with ever-bigger computer models when we know perfectly well that the small-scale stuff has no forecasting significance.

And indeed lots of people have tried to apply overall laws to very simple representations of the earth-atmosphere system. Effectively they assume that the very turbulence of the atmosphere and ocean on the smaller scales provides sufficient degrees of freedom for an overall law to be operative. All sorts of laws have been suggested. Some of them have been reasonably successful, at least for simulating present conditions, and some are even backed by real and provable physics. The problem is that it seems almost impossible to imagine ways in which broad overall laws can be applied to the present generation of climate models that are designed specifically to do their calculations on the small and so-called 'local' scale. The bottom line here may be a cultural one. The climatology profession is reluctant to give up on the hope that detailed forecasts are possible in principle. The prospect of having to put up with only the broadest averages is too difficult to countenance.

The point to all this, known at least subconsciously by almost everyone in the climate business, is that the fundamental problem of climate physics remains that of defining the minimum space and time scales over which it may be possible to predict changes in climate when somebody clouts Earth over the head with a sort of climatic bat.

We mentioned at the beginning of the chapter that there is a fair amount of reasonable science behind the global warming debate. True enough, but 'reasonable' is a relative term, and it has to be said that the typical climate model of today has great difficulty calculating even the present-day global-average temperature. The twenty or so models that have some respectability (by virtue of the fact that they figure largely in the IPCC deliberations) calculate global-average temperatures that range several degrees about the observed value of 15^0C. Their simulations of the broad *distribution* of present-day temperature are only so-so. And as might be expected from the earlier discussion, their simulations of the broad distribution of other parameters such as rainfall can only be described as terrible.

Perhaps we should put something of a scale to the word 'terrible'. Recently there was an examination at the Australian National University of the various simulations of rainfall round the world as produced by the IPCC models. Their simulations of average Australian rainfall ranged from less than 200 mm per year to greater than 1000 mm per year. The actual measured value is about 450 mm per year. More to the point, of the various *forecasts* by the models of late 21st century rainfall, rather more than half predict an increase over Australia, and somewhat less than half predict a decrease. The average is for an increase of about 8 mm per year. The most extreme of the forecasted decreases comes from the CSIRO model. (The Commonwealth Scientific and Industrial Research Organization is the major scientific agency of the Australian government). That model would have us believe that the average rainfall over Australia 100 years into the future will be 100 mm per year less than it is now. Which is a bit of a problem since it was this forecast that shaped the CSIRO input to the calculations of economic disaster that in 2008 became the basis of the proposed Australian response to global warming.

The most scientifically interesting result from the examination of the IPCC rainfall predictions is that some of the models give entirely different results if very small changes are made to the initial conditions. This evidence for chaotic behaviour associated with simulation of Earth's climate does not bode well for future improvement in the forecasting skill of climate models. Far-future regional rainfall may indeed be inherently unpredictable.

In terms of 'hind-casts', the Little Ice Age of the seventeenth and eighteenth centuries and the mediaeval warm period of the fourteenth and fifteenth centuries do not emerge from model simulations – at least not without a lot of very suspicious artificial help. In other words it is not known if even those very large-scale climatic events were caused by something identifiable or whether they were random and inherently unpredictable fluctuations emerging from the turbulence of the system.

Climate Averages on the Largest Scale

Enough of detail, let us look only at global-average climate, since we can be reasonably hopeful that averages on this scale may be predictable in principle. The specific concern in this section will be with 'feedbacks' on global-average temperature. A feedback occurs when a change of temperature from a particular cause changes something else in the system that in turn amplifies or reduces the original change in temperature. Physicists refer to the measure of the amplification or reduction as the 'gain' of the system.

A doubling of the concentration of carbon dioxide (CO_2) in the atmosphere will probably occur over the next hundred years or thereabouts because we are burning lots of fossil fuel. It is fairly easy to calculate the likely rise of global-average surface temperature caused by such a doubling, provided that we confine ourselves to the purely theoretical situation where nothing else is allowed to change. Such a calculation can be done reasonably easily these days (not quite on the back of an envelope but you get the drift), and was first done more than a century ago quite without the aid of number-crunching computers. The answer is just over one degree Celsius, and it would take two or three hundred years to complete the change. The problem is that in the real world there are lots of other things happening that can 'feed back' on surface temperature. Some of them amplify and some of them reduce any change caused by an increase of carbon dioxide.

The very first point to make is that, whatever are the feedbacks in the real world, and even if they were all negative in the sense that they would all reduce the basic change caused by the extra CO_2, the final temperature change is still going to be an increase. In this restricted sense the bald statement that "the science of greenhouse warming is proven" is indeed acceptable. Increasing CO_2 will certainly lead to higher temperatures than would have occurred otherwise. What is not acceptable is the corollary rather naughtily implied by those who loudly proclaim the statement about proven science in public – namely, that the rise in temperature has been proven to be so large as to be disastrous, or even so large as to be noticeable.

All the vastly complicated numerical climate models of today are machines which, in effect, calculate their own particular feedbacks by simulating as many as possible of the detailed processes of the earth-atmosphere system. The final forecast of each model depends largely on the particular set of tuneable parameters that has been chosen to make the model work. The range of predictions for the future is enormous, and it is not surprising that, more or less by chance, at least some of them are frightening. The best that scientists can do is to take some sort of average of the forecasts from those models deemed reasonably respectable.

Now taking an average may be sensible enough when dealing with experimental observations, but it is highly suspect when attempting to select one particular theory from a range of theories derived without the benefit of comparison with a controlled experiment. The significant word here is "controlled". It is true (well, it is probably true) that the earth's average temperature has risen by some fraction of a degree over the last hundred years, more-or-less in concert with a rise in concentration of atmospheric carbon dioxide that has (again probably) been caused by the burning of fossil fuels. This observational result is the only experimental support to which the models of global warming can refer in an attempt to verify their overall forecasting ability with regard to surface temperature. Since it is known that Earth's temperature has continually gone up and down like a yo-yo on all sorts of time-scales in the past and presumably will continue to do so in the future – and this for reasons as yet unknown – one can scarcely regard a particular century-long coincidence with increasing carbon dioxide as a controlled experiment.

Of course there are a lot of other experimental observations relevant to climate change, and we are forever hearing that glaciers have retreated, tree rings have thickened (or thinned), wind speeds have gone up (or down), rainfall has increased (or decreased) and so on, all of which are hailed as proofs of global warming. They may be support of a sort, but in fact all they support is the basic observation that the average temperature has gone up over the last century. They say nothing further about the reliability of climate forecasting or about the response of the earth-atmosphere system to increasing atmospheric carbon dioxide.

As a consequence, climate forecasting has become an exercise of inter-comparison of theoretical models rather than comparison with experiment. So let us look in a little detail at what the models actually say. We have to be a little technical and introduce a mathematical equation and a couple of graphs, but it isn't all that complicated.

Imagine that the basic rise without feedbacks of global temperature from doubled CO_2 is ΔT_0. Imagine as well that g_1, g_2, g_3 and so on are the actual values of the individual feedback 'gains' associated with each of the various atmospheric processes dependent on surface temperature. They may be positive or negative. That is, they may amplify or reduce the basic rise in temperature ΔT_0 associated the increase of CO_2. The total gain G of the overall system is simply the sum $(g_1 + g_2 + g_3 + \ldots)$ of all the individual gains, and the actual temperature rise ΔT when all the feedbacks are allowed to operate is simply the value of ΔT_0 divided by a factor $(1\text{-}G)$ as shown in the equation:

$$\Delta T = \frac{\Delta T_0}{(1 - G)}$$

The mathematically minded of you will recognize that all sorts of trouble would arise if the total gain G were 1.0. The actual temperature response would be infinite.

The diagram on the right is a graph of that relation between ΔT and G. Note the 1.2 degree Celsius rise for no feedbacks, the infinite rise for G equal to 1, and the edges of the cross-hatched area which indicate the range of total feedback gains and corresponding temperature rises for the dozen or so respectable models for which information on feedback is available. The range indicates that the total gain of an individual model falls somewhere roughly between 0.4 and 0.8. The corresponding range of temperature rise lies between 2 and 6 degrees Celsius.

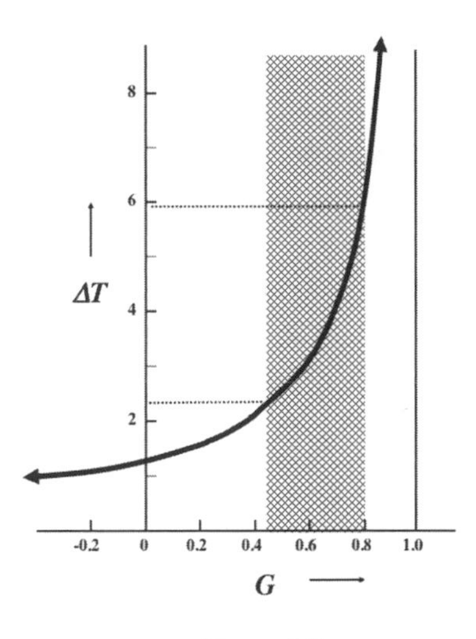

Figure 2.2

The definition of 'respectable' in the present context is a bit loose. It is close enough to saying that we are concerned with the set of complex numerical climate models that form the basis of the forecasts by the Intergovernmental Panel on Climate Change.

There are a couple of other things worth saying about the diagram.

First, it should be recognized that it refers to the temperature change that would be experienced once things have settled down after the doubling of CO_2. The response of any system to any change takes a little time to complete and, in the case of Earth's climate system, that time is of the order of a century or more – this mainly because of the vast amount of energy required to change the temperature of the oceans. So in the short term, let us say in less than 100 years, the real change of temperature would be rather less than might be indicated on the diagram.

Second, the shape of the graph is such that an increase of total feedback is far more effective in changing the temperature upwards than is an equivalent decrease of total feedback in changing the temperature downwards. Basically this means that it is far easier to make subconscious changes in mathematical representations within a climate model that produce frighteningly large temperature rises than it is to make changes that produce reductions. Put another way, the temperature rise produced as the average from a random selection of climate models will tend to be higher than one might normally imagine.

Let us be a little more complicated and look now at the individual feedback gains g_1 and g_2 and so on that are associated with each of the major feedback processes typically built into the respectable climate models. The next diagram (Figure 2.3) has vertical solid lines which indicate the range of each of the individual-process gains that can be found by looking over all the various models. Put another way, an individual model has a gain for a particular type of feedback which falls somewhere along the relevant vertical solid line. For those of you who are interested, the information is derived from a research paper by Sandrine Bony and many of her colleagues. It was published late in 2006 in the Journal of Climate.

The feedback processes are associated with the behaviour in response to temperature change of water vapour (*WV*), cloud (*Cl*), reflection (*Re*) of sunlight by the ground, and a thing called lapse rate (*LR*) which we won't bother explaining except to say that it concerns the rate at which temperature decreases with height in the atmosphere. Also on the figure is the spread over all the models of the overall gains *G*. It corresponds (as it should) to the edges of the cross-hatched area in the previous figure and ranges from about 0.4 to 0.8.

In principle there is no reason why any particular model's set of individual-process gains are more realistic than any other. So if all the individual processes were truly independent there should be no reason why the spread of total gain could not be as large as that indicated by the vertical dotted line on the diagram. That is, the overall gain *G* could range from less than zero to something greater than one. Suffice it to say that the relatively narrow range of total gains displayed by the actual models (roughly 0.4 to 0.8) is fairly surprising, and must come about for one of two reasons. Either the individual process gains are linked – 'correlated' as they say – in some convenient way, or there has been some subconscious choice of

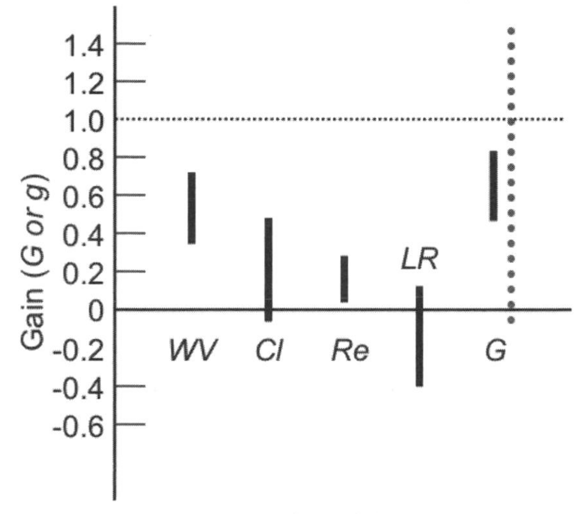

Figure 2.3

process description to keep the total gains of the various models within physically realistic bounds. One suspects that there is a bit of both involved in the business. On the one hand for instance, the research literature talks at some length about the correlation between water vapour and lapse rate. On the other, a lot of subconscious tuning effort goes into ensuring that climate models don't run off the rails of reasonableness.

Each of you can go back to your desk and build your own climate model based on the equation quoted above by picking any value you like for the feedback gain of each of the above processes. Provided your choices lie somewhere in the ranges indicated by the appropriate vertical solid lines of the diagram, your climate model will produce forecasts just as acceptable as those of anyone else. Well, perhaps that is going a bit far. Lets just say that it would be extremely difficult to prove you wrong.

The most uncertain of the individual feedback gains must be that to do with cloud. The range shown in the diagram is mostly positive, which in physical terms comes about because the models generally predict that higher temperatures are associated with less cloud. This is intuitively a bit suspicious since one might reasonably imagine that higher temperatures go with more water vapour in the atmosphere

– and therefore more cloud. Unfortunately the matter is completely unarguable since there is no direct experimental evidence one way or the other. Certainly a large negative cloud feedback (as likely a situation as any other) would drag the total feedback right down and lead to much smaller increases in temperature from increasing CO_2 than are currently fashionable.

And it is still quite possible that the apparently dominant positive feedback of water vapour is much less important than any of the respectable models suggest. Indeed there is experimental evidence from balloon measurements over the last 50 years to support the possibility that water vapour feedback is actually negative, although satellite measurements generally do not. Both types of measurement have their problems, and at this time it is largely a matter of personal preference as to which source of information is the more believable.

The summary of all this is that climate modelling cannot really be described as robust when it wouldn't take much fiddling with the individual feedback representations to give temperature rises covering the whole range from much less than 1 degree Celsius to infinity and beyond. And to make matters worse, there is nothing holy about the physics of the respectable models. The real feedback gains could be way outside the range of the current representations. Put most generally, one of the great myths about global warming is that the science is settled.

It is a great mystery as to what might be done about the matter, since there is precious little direct experimental verification of any of the process representations. The big danger is that, with increasing model complexity and cost, the number of truly independent climate models around the world is decreasing. This is because great slabs of the computer code of a model are often exchanged between research groups so as to avoid writing the stuff from scratch. This sort of exchange satisfies a general bureaucratic tendency to abhor what seems to be duplication of effort. The net result must surely be a natural decrease in the spread of total feedback over the various remaining models, and a consequent joy at the apparent tightening of the range of forecast temperature rise – a tightening that may have nothing at all to do with an improvement in the representation of the physics.

One ought to be able to solve that particular bureaucratic problem, but the politics of the situation suggest that it is not likely to happen any time soon.

As a final random thought, it is at least theoretically conceivable that the total feedback gain of the climate system is actually very close to 1.0. In such a circumstance one could imagine the climate skating from one extreme of temperature to another and back again. The extremes would be the points at which the total feedback gain became less than 1.0 – as for instance when cloud cover reached zero or 100 percent and could no longer contribute to the feedback. After all, the climate has always been flipping in and out of ice-ages! More to the present point, and were such a situation to exist, it wouldn't matter very much whether or not man added lots more carbon dioxide to his atmosphere.

Why Are Things So Uncertain

Still sticking with the broad issue of forecasting global-average climate, it is worth spending a little time discussing just why there is so much uncertainty about the business. Why is it that the actual values of the basic feedbacks are so questionable?

The most general answer is that there is no experimental information that can be used to 'tie down' the values of those feedbacks contributing most to amplification of the rise of Earth's surface temperature in climate models. To be strictly correct, we should say that there is no experimental information accurate enough for the purpose. Furthermore, it may be some considerable time before we get it. If ever. The main problems are the feedbacks associated with cloud and water vapour in the atmosphere.

The rest of this section is a little technical, so the fast readers among you may care to give it a miss. It describes with hand-waving argument a bit of the physics behind the statements of the previous paragraph.

Earth is hot because it absorbs a vast amount of solar energy. Because it is hot, it radiates infrared energy back to space. Because the amount of radiated infrared energy increases with temperature, and because in the end the overall system must settle down to a sort of steady state where things don't change much over the long term, the earth must adopt a temperature (and a temperature distribution) that ensures there is a close balance between absorption of solar energy and emission of infrared radiation. The major twist to the picture is that the infrared energy radiated to space is emitted both from the earth's surface and from the tops of the 'blankets' of carbon dioxide, water vapour and cloud in the atmosphere. The temperatures of the blanket tops are much less than that of the surface because they are high in the atmosphere.

If we increase the concentration of atmospheric carbon dioxide, we effectively increase the thickness of the CO_2 blanket. The top of the blanket becomes cooler because it is higher, and as a consequence it radiates less infrared energy to space. Since the amount of absorbed solar energy is more-or-less fixed if nothing else is happening at the time, infrared radiation from the earth's surface must increase to ensure that the total energy radiated to space continues to balance the solar energy input. In other words the earth's surface temperature must rise.

Note that it is what happens at the *top* of the blanket that governs the temperature rise at the surface.

The same story applies to water vapour and water vapour feedback. If increasing CO_2 raises the surface temperature, then indeed one might reasonably expect the rate of evaporation of water from the surface to increase. (Actually this is not a certainty, but it is probably true on average). The global rainfall will increase because it must balance the evaporation, and in the process the amount of water vapour in the atmosphere will increase. (This is not a certainty either, but it seems reasonable). The increase in water vapour will in turn, if one says it quickly, increase the thickness of the water vapour blanket and will further increase the surface temperature for much the same reason as with increasing CO_2.

Let us be a little more careful with that last sentence.

Again it is what happens at the *top* of the blanket that matters. Provided the water vapour concentration goes up more or less proportionally at all heights in the atmosphere as the surface temperature rises, then indeed the water vapour feedback will be positive. It will amplify the temperature rise due to CO_2 alone. The computer models of climate behave in just this way. On the other hand, it is perfectly possible that in the real world the small water vapour concentrations in the upper levels of the atmosphere – above about 3 or 4 km or thereabouts – could decrease as the larger concentrations in the lower levels increase. There are plausible physical reasons why this might be so. In which case the overall water vapour feedback would be negative and the original temperature rise from increasing CO_2 would be reduced.

If the reader is wondering why we are making rather a meal of this particular discussion, one of the reasons is that it is not well known even among the cognoscenti of the climate modelling fraternity that it is the behaviour of water vapour in the upper atmosphere (to be semantically correct, in the middle and upper troposphere) which determines whether water vapour feedback is positive or negative. And it is at just these levels where the world's fifty-year-long record of balloon measurements of atmospheric water vapour (the record shows a decrease of water vapour over the years) is the most doubtful. Some would say it is probably nonsense. It is at just these levels also that the world's 25-year-long record of satellite remote measurements (which show an increase of water vapour over part of the time) is the most doubtful. Some would say that it too is probably nonsense. Suffice it to say that at this time there are no experimental data with which to settle the question about whether water vapour feedback is positive or negative.

Feedback associated with variation in cloud amount is even more of a problem. Way back in the early days of formalized climate research – that is, back in the early 1970s when the World Meteorological Organization in Geneva was attempting to define what needed to be done in the way of climate research – it was recognized that the biggest problem was how to simulate clouds in climate forecasting models. A formal program (the International Satellite Cloud Climatology Program) was designed and developed specifically to monitor the amount and type of cloud about the world. It has provided much interesting information. It has established for instance that average global cloud cover is nearer 70 percent than the 50 percent assumed as a rough working number in earlier times. However it has a long way to go before it can hope to obtain measurements of cloud and cloud change sufficiently accurate to resolve the question as to whether cloud feedback is positive or negative.

The scientific difficulty is that cloud reflects incoming solar radiation back to space and therefore tends to cool the world – this as well as behaving rather like the water vapour and carbon dioxide infrared blankets and tending to warm the earth because of its effect on outgoing infrared radiation. Which of the processes 'wins' – that is, whether the overall effect of cloud is a cooling or a warming – depends on both the height and the character of the cloud concerned. And again as for CO_2 and water vapour, whether overall cloud feedback is positive or negative is very largely dependent on the exact height, character and distribution of cloud in the upper levels of the atmosphere. Just those levels in fact where satellite measurements are hard put to it to make sufficiently accurate measurements. New satellites with new instruments that may help solve the problem are being launched, but it will be many years before they produce sufficiently accurate data to answer questions about the possible change of cloud and of cloud character over time.

The Problem of Climate Model Complexity

In pursuit of ever-more-detailed description of atmosphere and ocean in an attempt to reduce the need for tuneable parameters and to provide more realistic representations of the various feedback processes, climate models expand to fill the computer space available. They are run on the biggest computers in the world, and may need months or years of computer time simply to settle down into a sort of equilibrium state before an actual trial can begin. Any particular trial is extremely expensive and can take years to complete. As a consequence there is neither the time nor the inclination to run basic tests of the sensitivity of a forecast to a change of any of the tuneable parameters. It is far easier simply to ensure alignment with the corresponding parameters of other models.

And because the models are so complex, it is very rare for a group of researchers to develop a new climate model from scratch. They take some or all of the code from the model of another group, and slightly modify those bits of it that are relevant to their particular interest and expertise. The overall process ensures that there is a gradual, and largely unconscious, move towards a situation where all the supposedly independent models have common physics and common values for their tuneable parameters. Quite naturally they begin to tell the same story.

The danger is that the narrowing range of answers will be interpreted by scientists – and particularly by their managers – as an indication that the accuracy of their climate forecasts is improving.

No scientist outside the closed engineering shop of the numerical modelling community can ever really hope to assess whether or not the physical representations within them are acceptable. Quite apart from anything else, how can an outsider lay his hands on the team of numerical engineers necessary to run, modify and generally understand the way a particular model is behaving? More important, how can a reviewer of a research paper reporting results from the trials of a particular model make any assessment of those results other than to base his opinion on whether or not they fall in line with the results of other models?

This is the core of the problem. Over the last few years a number of scientific articles have emerged which compare the basic controls of climate change – the feedback parameters if you wish – calculated by the various models. The papers provide the average values and the spread of these parameters, and as a consequence it is now relatively easy to see whether one's own particular model is behaving 'normally' – that is, more or less the same as everyone else's model. The subconscious pressure for teams and organizations, to say nothing of individuals, to conform to the average is enormous. There is no reason why a team should go to the trouble of pursuing a line of research based on feedback parameters that lie outside the expected range. Reviewers will simply say, because there is precious little else on which they can comment sensibly, that the results don't match the consensus.

In short, the normal and necessary process of scientific criticism cannot take place. All that can happen is for sceptics to take verbal pot-shots at the modellers purely on the basis of intuition, and for the modellers to defend themselves by saying that no-one outside the modelling game has the knowledge to make sensible comment.

It is not the way to run a scientific railroad. We have to get away from simply running models and comparing their final output in some sort of search for a consensus on the results. Consensus is not science. Consensus tends to the politically correct. Consensus is not the sort of thing on which sensible people put their money. Not in the climate game anyway.

Perhaps one of the biggest problems with numerical modelling is that it is a gentlemanly activity conducted entirely from one's desk. There is no need to visit the real world too often. In compensation, each run of a theoretical climate model is routinely referred to as an 'experiment' and one can conduct one's 'experiments' with lightning speed. Productivity in terms of results and publication is enormous. The fact that the results may have nothing to do with observations from the real world can conveniently be ignored. Other models take the place of observation.

The Bottom Line

One of the more frightening statements about global warming to be heard now from the corridors of power is that "the scientists have spoken". Well maybe they have - some of them anyway - but the implication of god-like infallibility is a bit hard to take.

The science of disastrous global warming is far less settled than climate activists would have us believe. The high probability attached by the IPCC to its thesis of climatic disaster is not the result of careful scientific analysis of theory versus experiment. Basically it derives from a set of people sitting round a table making personal guesses about thc quality of the models. One suspects they ignored the possibility that, because the bandwagon is rolling so well, there is very little chance these days that a model will ever be allowed to stray too far from the average.

Such a statement needs a little explanation. Since experimental data simply don't exist to check the model calculations of an amplified temperature rise, so also there are no experimental data to support any 'outlier' model whose owners are foolish enough to calculate a global warming that is not really all that serious. The existing models were established in the scientific literature well before the political stakes got so high, so they hold the high moral ground. Getting an 'outlier' into the system is almost impossible because the would-be author of dissenting results is expected to prove why his calculations are better than those of other models. And of course he can't.

CHAPTER 3

SOME ECONOMICS

"Economist Ross Garnaut tells us that, by the year 2100, all the normal measures of Australian standard of living (real wages and per capita consumption and so on) will be somewhere between five and ten percent below what they would be if humans were not filling the atmosphere with carbon dioxide. He also tells us that, by 2100, and in the absence of global warming, the Australian economic output per person would increase by about 400 percent. Provided the government didn't waste it, most of that would come back to us in the form of increased standard of living. Therefore the average Australian of 2100 should be something of the order of 360 percent better off than ourselves even when he or she has been devastated by climate change. In other words Professor Garnaut is asking for lots of our money so that he can give it to people of the future who will be at least three-and-a-half times wealthier than we are. The guy must be nuts."

Letter to the Australian Financial Review, July, 2008

There is indeed a measure of resigned acceptance these days within the ranks of the movers and shakers of many countries that Earth will warm significantly over the next century because of extra input to the atmosphere of carbon dioxide and other greenhouse gases derived from human activity. Or perhaps it would be more accurate to say there is a measure of resigned acceptance that many people these days believe the proposition. As a consequence either way, the nations of the world are actively negotiating with each other in an attempt to limit their emission of greenhouse gases. The object is to slow or halt the change of climate that may follow as a consequence of the 'enhanced greenhouse effect'. This at least is the object as proclaimed loudly by the participants. There are lots of other agendas attached to the business.

Any normal man or woman in the street would imagine that the negotiating process involves a quantitative analysis by each nation of the present costs and future benefits to itself of whatever is the set of actions proposed by the international community. Unfortunately their imaginings are a bit wide of the mark. While it may be possible to estimate the present costs of a particular action, it is virtually impossible to say anything quantitative, or indeed anything sensible, about the benefits deriving specifically from the reduction of the future impact of climate change. Bear in mind that 'the future' in this context is something of the order of 100 years from now.

We made the point earlier that scientists cannot yet translate, and in fact may never be able to translate, their general prediction of an overall warming of the earth to forecasts of the specific change of climate which might occur in any particular region. Therefore economists have no solid input information with which to drive their own arcane models of a regional or national economy. In any event, and even if the climatic input information were accurate, economists are hard put to it to predict the likely impact on a region of any particular change of climate. This for a number of reasons, the most cogent of them being that the world economy has many of the same turbulent characteristics as those of the atmosphere and ocean. In particular, while the broad behaviour of the global economy might be predictable in principle, if not in practice, it is very possible that the behaviour of an individual national or regional economy is inherently unpredictable beyond a few years into the future.

Indeed, the forecasting problem is much worse for an economist than a climatologist. The physical climate system is at least very close to what the scientists call 'steady state', which means that its overall characteristics remain fairly constant with time, and its input (solar energy) is closely balanced by its output (infrared radiant energy back to space). Steady state systems are vastly easier for scientists to understand and to handle mathematically than are non-steady-state systems whose behaviour often tends to be counter-intuitive and difficult to visualize. It is not for nothing that development of an understanding of non-steady-state systems is one of the great unsolved problems of science. It is on a par in many ways with the search for a unified field theory, or with the struggle to understand the origin of life.

Suffice it to say that the economies of the world and of the various nations are systems far from steady state. They are continually increasing in overall size and are changing in character – well, we hope they are continually increasing in size – and there is usually no balance between whatever are their inputs with whatever are their outputs. Their behaviour is often counter-intuitive, and is often difficult to understand even after an event. All in all, the potential for inherent unpredictability in an economic system is vastly greater than it is for the mere physical system of Earth's climate. It is not surprising that the reputation of economists and of their long-range forecasts leaves a lot to be desired.

And for many of the same sorts of reasons, as well as for a number of reasons unique to their disciplines, ecologists and biologists cannot yet predict the likely impact of a specific change of climate on the natural environment.

In short, the problem of calculating the long-term benefit of an expensive exercise to prevent climate change has every chance of being inherently insoluble.

The Politician's Particular Problems

There are many purely political difficulties associated with assessing the benefit of active limitation of climate change. They arise because the impact of climate change may be detrimental or beneficial to a country as a whole; because some sectors of its economy may gain and others lose; because in the global context some countries may gain and others lose; because the cost of a particular strategy for prevention of greenhouse gas emission will inevitably vary greatly from country to country; and because the cost and success of a prevention strategy in one country will generally depend on the prevention strategies of others.

And just to make things a little more difficult for politicians, they have to deal with two very fundamental problems which, whether they realize it or not, lie behind most of their decisions on virtually any matter of broad social interest.

The first is that it is usually impossible to optimize simultaneously two or more characteristics of an overall system. Normally for instance it is a nonsense to suggest, as many do when trying to sell a course of action to the public, that one can arrange things to achieve both maximum benefit and least cost. One cannot adjust the actions of nations in such a way as to maximize the benefit to cost ratio for each of them at the same time. One cannot expect to maximize the ratio for a single country while maximizing the ratio for all its internal sectors. One cannot expect to maximize simultaneously both a benefit to the present generation and a benefit to generations of the future. And certainly one cannot expect to maximize simultaneously the value of two sectors of society when the units of 'value' of one are entirely different to those of the other. Choices in such cases can only be made by reference to one's gut feeling of the present mood and beliefs of the population to which one belongs.

The second problem is related to the first and concerns the actual definitions of 'cost' and 'benefit'. The basic problem is to assign a value to those environmental aspects of concern to society – things like biodiversity, environmental quality, health and so on – which do not fall within the normal economic definitions of national worth. In particular the problem is to define an appropriate 'discount for the future' for such things. It is difficult enough to assign a dollar value to the biodiversity of the world as it is at the moment. It is even more difficult to obtain agreement on the value *to us now* of the biodiversity of the future. Many of the most powerful arguments for limiting greenhouse warming are based on preserving the current state of biodiversity and of the environment. All of them imply a discount for the future that is very small compared with the discount appropriate to the purely economic sectors of society. The choice of a particular environmental discount for the future has to be made – even if only subconsciously – if costs and benefits associated with responses to greenhouse warming are to be assessed. Inevitably the choice is a value judgment. Value judgments are the very stuff of politics. In fact the need to make such judgments is precisely why we pay politicians. They are expected to make decisions on matters for which there can be no solid evidence one way or another.

Perhaps we should expand a bit on this 'discount for the future' business. The easiest way to think about it is to ask oneself a series of questions beginning with "how much would I be willing to pay today in order to provide a benefit to people living one million years in the future?" There are very few people indeed who would put good hard-earned present dollars into such a venture. Quite apart from anything else, there is a distinct possibility that there would be no people around at that time to benefit from our largesse. So what then would be the point?

Let us shorten the time scale a little, and ask the same question with regard to people living one thousand years in the future. Some of us might be tempted to spend a bit on such a cause, but most of us I suspect would look back at the changes that have occurred over the last 1000 years and wonder whether anything we do now will be significant in the context of the changed society of 1000 years ahead. Most of us would keep our hands in our pockets.

If the time scale is shortened to only one hundred years, the question becomes much more relevant and sensible. Quite a few of us, and particularly if we were reasonably wealthy, would be willing to pay out a bit of money to provide a benefit to people three or four generations ahead. After all, while we might not know these particular future people personally, with a bit of luck we might know someone who would.

And so it goes. The nearer a potential benefit is to the present, the more we are willing to pay for it.

For purely economic things – that is, the sorts of things we are accustomed to valuing in dollars and can in a broad sense buy off-the-shelf – the general belief is that their value to us now will reduce more-or-less to zero if we were buying them for the benefit of people more than a generation or two down the line. Put another way, the rate of future discount for such things is at least of the order of several percent per year. For environmental benefits – that is, the sorts of benefits to which we cannot easily assign a value in dollars because our interest in them is based on our personal judgment of the need to preserve the status quo – the rate of discount for the future is generally much smaller. Give or take a bit, it is probably something of the order of a percent per year, which roughly corresponds to a practical concern with the environmental welfare of future people into the third or fourth generation.

In summary of all this, in order to make any reasonable assessment of future benefit associated with an action which has a current cost, it is an absolute necessity to decide on some particular rate of

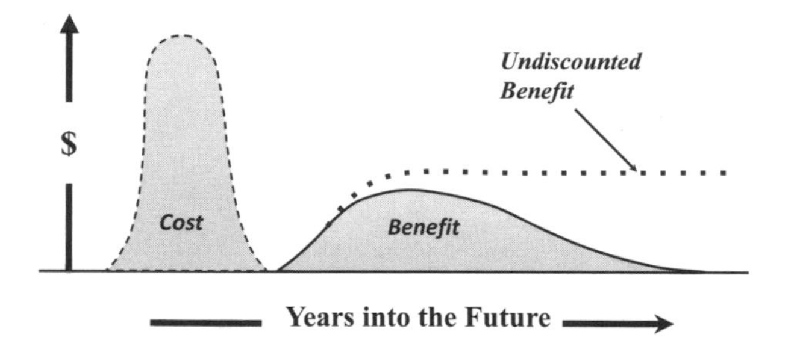

Figure 3.1

discount for the future whether it be for purely economic goods or for the more esoteric environmental goods. Thinking about Figure 3.1 makes this clear enough. The figure shows the hypothetical cost per year of some reasonably immediate action (the dashed line), the future benefit per year resulting from that action if some discount for the future is applied (the solid line), and the benefit per year if the concept of future discount is ignored or is zero (the dotted line).

For those of you familiar with a little mathematics, the total cost of the action is the shaded area under the dashed line. The total benefit when a future discount is applied is the shaded area under the solid line. And for that situation the benefit-to-cost ratio is simply the ratio of the two areas. If the ratio were less than 1, then most people would say that the action would not be worth taking since the return to them would be less than the cost. If it were greater than

1, then maybe it would be worth the trouble. The less the rate of future discount, the more likely it is that taking the action would be sensible.

But we have to be very careful. Extreme environmentalists would have us believe that there should be no discount for maintenance into the indefinite future of current biodiversity and current environmental quality. Which sounds good in principle except that the philosophy attaches an infinite value to the benefit of maintaining conditions as they are. Referring to the figure, the dotted line corresponding to zero discount extends out to infinity. In such circumstances the benefit-to-cost ratio is infinite, and we would be more-or-less obliged to devote all our current money and all our current resources to that purpose and to that purpose alone. Following such a course would demand a religious fervour far beyond the experience of normal human beings. It would also require us to believe that our current environment is the best of all possible environments.

When one thinks about this concept of discount for the future, it is obvious why climate activists are desperate to convince us that climate change will have an extreme negative impact within the next hundred years, and probably (preferably?) within the next few decades. There aren't too many of us who care a lot about spending current money for the benefit of people living more than a century down the track.

When one thinks about the concept even further, it is obvious as well why there isn't much mention around the traps about the very real possibility that the forecasts of future climatic disaster are wrong. The average person, and he is normally very sensible, would greatly increase his personal discount for the future if he were told that there is a significant chance that his worries about that future are groundless.

And to remind you what this rather esoteric discussion is all about, it is to make the point that a decision on how much effort we should put into reducing global warming – or indeed on whether we should reduce global warming at all – must inevitably be based on purely political considerations. In these days of political correctness, very few people would care to 'go public' with a view on what might be an appropriate discount rate to apply to the environment.

Inevitably the decision will be left to politicians. They, the poor souls, will be expected to determine by some magical means the innermost thoughts of their constituents on how much sleep we should lose over the plight of far-distant future generations.

The Modeling Business

The question each nation must ask is whether the potential cost of its own actions to reduce greenhouse warming can be justified when compared with the long-term benefit associated with a reduced impact of climate change well into the next century. Since the question and its answers are so much a matter of politics, it is not really necessary to perform highly complex analyses that demand detailed numerical calculations on some vast number-crunching computer. In principle the analyses need be no more complex than is compatible with back-of-the-envelope calculations and with the normal process of making political decisions.

If, despite all the uncertainties, one insists on making use of economic models to do cost-benefit analyses with regard to the climate change problem, then perhaps the most important characteristic of such models should be that they are simple enough for their basic assumptions to be challenged not only by numerical modelers, but also by the public. This is precisely the characteristic which is lacking in the current efforts being used to guide the political decision making process. In Australia for instance, the most that the man-in-the-street is told is that the advice to government is based on 'the Treasury model of the economy'. It is doubtful whether even trained economists outside the Treasury know how the thing works, let alone anything about its assumptions or whether its results for a hundred years into the future are anything better than guesswork. Come to think of it, most climatologists are in the same position with regard to the forecasting models of their own profession.

Nevertheless, let us spend a little time considering what might be the framework of an overall model concerned with assessing the costs and benefits of national actions to mitigate global warming. We will use it only as a vehicle to discuss a number of specific issues concerning potential responses to climate change.

Figure 3.2 is a cartoon representing an imaginary nation called Australia whose climate-sensitive activities are generating money each year and tipping it all into a bucket labeled Gross National Value (GNV for short). The cartoon imagines that the people of the nation – in fact the people of all the nations – are of saintly disposition because they do not indulge in activities which might increase the carbon dioxide concentration of the world's atmosphere. Every year the GNV bucket is almost full, and the money can be spent on vast numbers of desirable things like golf clubs and flat-screen televisions.

Figure 3.2

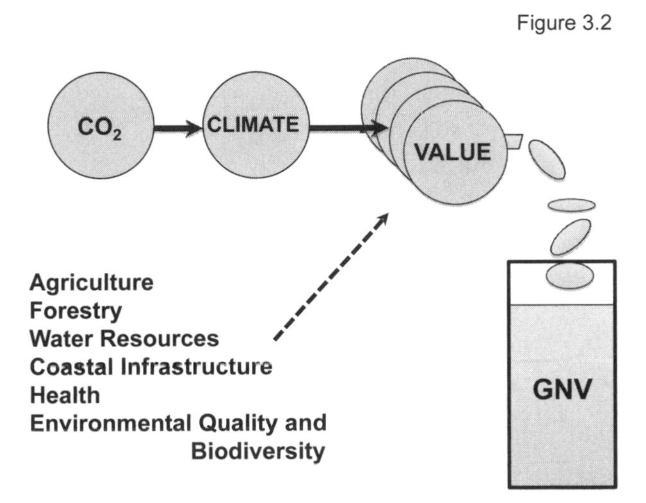

The main climate-sensitive activities have to do with agriculture, forestry, water resources, coastal infrastructure, health and the environment. The first two of these are purely economic sectors in the sense that the annual value of their production is easily definable in dollars and is easily counted as a contribution to gross national product. And the very first point one can make about these two sectors is that they don't necessarily make up a very large fraction of the total economy of a nation. This is certainly true of the larger well-developed countries. In the US and Australia for instance, the contribution of agriculture and forestry to gross national product is not much more than 5 percent, and is falling as the years go by. Which means that Professor Garnaut, who advises the Australian government on the impact of climate change and has talked about a 5 to 10 percent reduction in standard of living caused mainly by lower Australian rainfall of the future, must be envisaging an almost total wipe-out of that country's agriculture. An unlikely outcome in view of the proven adaptability of farmers, but there you are.

Figure 3.3 is an extension of Figure 3.2. The saintly people of the nation, along with the saintly people of all the other nations, have become rather more realistic and are burning lots of fossil fuel and thereby adding extra carbon dioxide (CO_2) to the atmosphere. Earth warms and its climate changes, and as a consequence the amount of money pouring annually into the GNV bucket is sadly reduced. Other nations are affected as well, so the balance of import and export of climate-sensitive commodities – basically of agricultural products – from one country to another is upset.

The impact on any particular country of an upset in international trade must be one of the most unpredictable things in the whole global warming business. With or without global warming, and with or without the help of enormous computer models, no economist in his right mind would bet on the shape of global trade one hundred years into the future. The changes wrought by normal development of human endeavour would vastly outweigh any change attributable to global warming. The unpredictability of future trade makes detailed modeling of the far future of any particular national economy little more than an academic exercise.

Also on Figure 3.3 is a pair of arrows meant to represent the need to take into account the periods of time required for changes to take place. The periods are called 'time constants' in the vernacular of the scientist. There are time constants associated with how long it takes particular aspects of climate to respond to a change in atmospheric carbon dioxide. In turn there are time constants associated with how long it takes particular sectors of the economy to adapt to a change in climate.

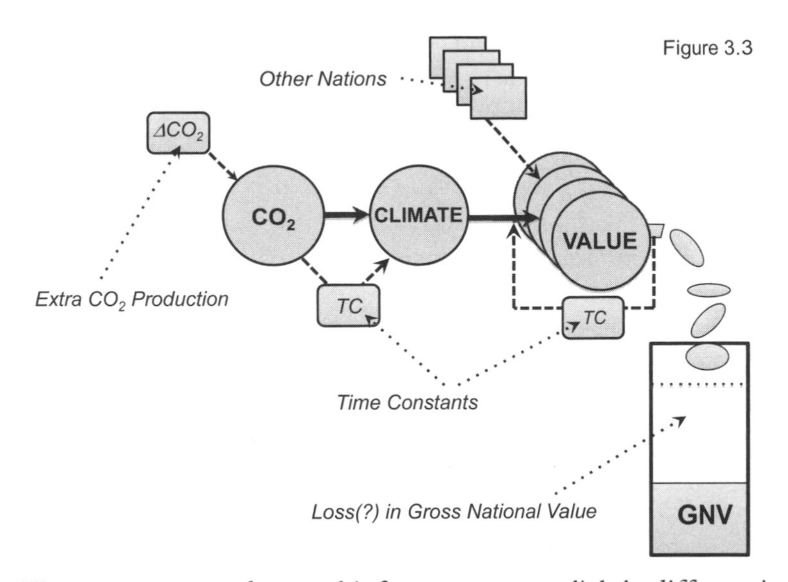

Water resources and coastal infrastructure are slightly different in that they don't provide a direct input to what we are calling the annual gross national value. Rather, they are background resources that must be protected against the impact of a change in climate. In their case the benefit of preventative action to limit CO_2 emission lies in the reduction of the future costs of their protection. That is, the benefit is a reduction in the cost of adaption.

In the case of water resources, the protection would normally be against a decrease in rainfall. As discussed in the previous chapter, it is possible that future rainfall may decrease in certain areas, but for most countries one might expect an increase since the overall global rainfall should increase in a warmer world. In the case of coastal infrastructure, the protection would be against a rise in sea level. Bearing in mind that current estimates envisage a rise of less than a few tens of centimeters over the next century (less than a foot in the old imperial measure) it is difficult to get too excited about the matter. The multi-century time constants associated with heating the deep ocean and with melting the polar ice caps ensure that really drastic changes of sea level, if they occur at all, will occur well beyond the time scale of typical human concern with the welfare of future generations.

As a general rule, if the time taken for natural adaption by an economic sector to climate change is much less than the time taken for the climate change to occur in the first place, then that sector will probably not notice much of an impact on its contribution to the national welfare. The dominant response time of the climate system to an imposed change of conditions is of the order of a hundred years. Agriculture, which is presumably the major climate-sensitive activity of humankind, has considerable adaptive ability. In countries like Australia and the US for instance, it is said that the time-scale for *active* conscious adaption to an agricultural disaster is of the order of a few years. The time-scale for natural (more-or-less unconscious and automatic) adaption to a change in the climatic environment is of the order of a single human generation. Suffice it to say that the typical farmer is not a fool, and is continually adapting his practices to keep up with changes of any kind. In particular he is continually and actively shifting his practices to take account of any change in the patterns of local weather. He may not like having to do it, but he is more-or-less used to it.

Health, and in a somewhat different way, biodiversity and environmental quality, are extremely difficult to quantify as economic goods. They are even more difficult to quantify in the annual terms that relate them to such things as gross national product or to the amount of money in the imaginary gross-national-value bucket of Figures 3.2 and 3.3. It seems that the only practical way to assign an annual value to biodiversity and to environmental quality is to consider the amount of money a national population might be prepared to pay to preserve them as they are. At the best of times this sort of consideration would have to involve political guesswork. The value would no doubt vary widely from generation to generation. It would also depend very largely on the wealth of the country concerned. It is reasonable to expect that poor people and poor countries will not entertain as great a concern for the long-term future of the environment as would people and countries that have a lot of disposable income.

If decisions eventually have to be made on such matters, it is probably as good a guess as any to say that a nation might be prepared to spend annually as much on preservation of the environment as it does on maintaining and improving human health. Whatever is that amount when expressed in terms of a percentage of the average person's income, we might well assume that the percentage would remain constant with time. Then it would be a matter of discounting that amount for the future in order to establish the value to us now of the environment of the future. Well, that's one way of looking at the issue. There are lots of practical and philosophical difficulties with such a method of assessment, but beggars can't be choosers.

Figure 3.4 is a final extension to the original Figure 3.2 situation. It envisages the present population attempting to restore its original saintliness by pouring money into measures designed to reduce emissions of carbon dioxide and thereby limit the degree of future global warming. It envisages also that future generations put their money into active adaption to any change of climate as it occurs. Either way (one might hope) the net result is a restoration of money in the bin of gross national value.

We can make two general points at this stage. First, the effectiveness of a national attempt to reduce its emission of CO_2 is completely

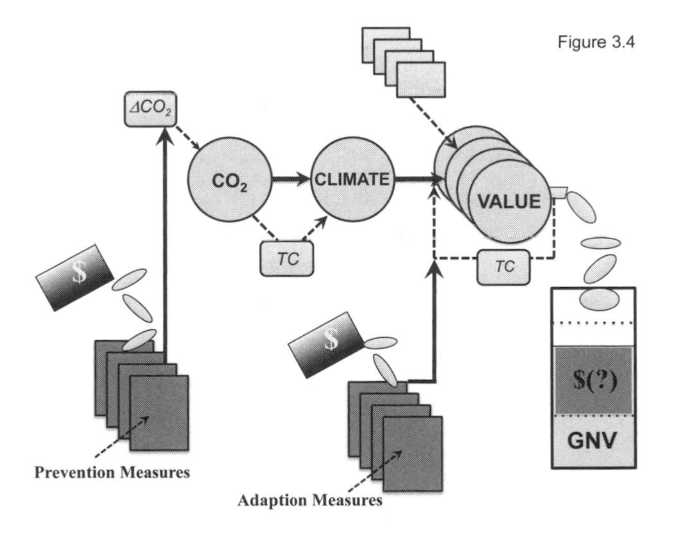

dependent on the prevention measures of other nations – this because all such measures make their impact via reduction of CO_2 in the common atmosphere. And incidentally, the costs of reduction must be paid in undiscounted present dollars. Second, the effectiveness of adaption is more-or-less independent of the actions of other nations. Therefore, and because he will have a relatively immediate benefit from his adaption, the man of the future will be reasonably certain of the benefit to cost ratio of his action before he starts, and will be able to make an uncomplicated decision. And from our point of view in the present, he will in any event be spending discounted dollars.

Short of closing down all the energy consuming activities of humankind, the options for reducing emissions of CO_2 seem to boil down to the following:

- conversion of existing coal-burning power stations to nuclear, gas or 'alternative energy' generators
- improving the efficiency of transport, or using an alternative fuel to drive it
- growing new forests, fertilizing the oceans, or actively burying the carbon being emitted by power stations - all three of which boil down to methods of disposing CO_2 someplace other than in the atmosphere
- reducing population growth

On Power Stations

The first of the options is a sensible-enough strategy provided that it is done over the full lifetime of the existing coal-burning infrastructure. An insistence on immediate replacement of current power stations would simply ensure that the nation throws straight down the drain an enormous number of current dollars invested in current capital. Given enough time, an existing power station will need to be replaced for other reasons, and one could reasonably ask the industry to change to another form of generation at that time. Of course it might be that the other form of generation would produce more expensive electricity, but to some extent the future discount of that extra expense would enter into the cost-benefit calculations of today's planners.

This raises an issue that has again to do with the actual science behind the global warming debate. One of the loudest mantras of the committed climate-change activist is that we are running out of time to do something about the problem – that in another very few years it will be too late to do anything at all because we will have reached irreversible 'trigger points' which will signal the sudden onset of Armageddon. The scientific background to this sort of statement comes in part from what is called the 'non-linearity' of certain elements of the physical climate system. (Non-linearity raises the possibility that one can give a small push to one part of the system and there may be a very large response somewhere else). This sort of phenomenon is part and parcel of the reason for the unpredictability of climate on the smaller scales.

Instability as a consequence of non-linearity is less and less likely in a turbulent atmosphere and ocean as the size of the area of concern increases towards that of the total globe. The way the scientists look at it is that, with increasing size, there is an increase in the number of 'degrees of freedom' in the system. More degrees of freedom should allow the broad overall behaviour of the system to be more linear and predictable. Certainly the atmosphere-ocean system as a whole seems to survive the shock of seasonal changes without going berserk. That is, there is no evidence of an unstable response to the massive re-distribution of the input of solar energy that occurs between summer and winter.

However, the concept at the back of the mind of most people who seize upon the trigger-point concept as a reason for immediate action is probably that carbon dioxide in the atmosphere is increasing faster and faster as the years go by. The phenomenon sounds more frightening when it is expressed in mathematical terms - namely, that the increase of CO_2 is 'exponential'. Either way, the idea is that pretty soon the increase, and the consequent change of climate, will be so fast as to be unstoppable. The problem with the idea is that the actual impact of more CO_2 on the world's temperature becomes less as the amount of CO_2 in the atmosphere increases. In mathematical terms, the response of the world's temperature to increasing CO_2 is 'logarithmic'.

To put the reader out of his mathematical misery, the bottom line here is that the combination of an exponential increase of CO_2 with a logarithmic response of temperature means that the rate of change of temperature and climate will be more-or-less constant with time. On that basis, and other things being equal, the change over the next 50 years should be no worse than the change over the last 50 years. Which suggests in turn that there is no inherent need to rush into the business of limiting the human emissions of greenhouse gas. Things are not likely to get out of hand all of a sudden.

Finally on the specific topic of converting power stations, on the face of it the very worst way of encouraging the industry to do something about conversion would be to subject the industry to the sort of trading scheme that would deplete the store of capital ultimately available to fund the changeover. In many ways such a scheme would be even sillier than forcing an immediate conversion, since among other things it would inevitably divert accumulating capital into the unlimited pockets set aside for other purposes of government.

On Transport

Improving the efficiency of transport, or replacing petrol and diesel with an alternative fuel, would seem to be reasonably sensible responses to global warming simply because they are the sorts of thing which nations may have to do eventually anyway as global sources of easily-accessible oil dry up. The only moot point is whether the changes should be massively forced and supported by government or whether the world should wait until a natural increase in the price of oil forces the change of itself. Bearing in mind the political difficulties associated with the rather curious distribution of the raw material about the world, it is understandable that the option of massive government support will probably win the day. While there is no obvious hurry to do anything drastic in the global warming context, there may be an immediate need to do something drastic if the major oil supplier nations play too dangerous a game of geopolitics.

On the Disposal (Sequestering) of Carbon Dioxide

The various options for disposing CO_2 somewhere other than in the atmosphere raise some peculiar issues.

Growing more forest is a method of salting away the carbon of atmospheric carbon dioxide into wood. It is a good idea from a number of points of view, and in particular it is pleasing to the modern environmentalist who is rather keen on trees. One has to bear in mind that it is a 'one off' solution in the context of limiting the increase of atmospheric CO_2. That is, the benefit continues only while the forest is actively growing. Leaving the trees to die, fall over and to rot as in the natural cycle of things would simply ensure that the captured carbon eventually returns to the atmosphere. The situation would be as before. In order for there to be a long-term benefit, the trees would need to be buried – and buried fairly deeply at that. More sensibly perhaps, they could be used as construction timber and thereby would lock away their carbon for a much longer time.

There is an interesting downside to the tree-growing option. Land covered with forest is rather dark compared to most other surfaces. Its 'albedo', the fraction of down-coming solar energy it reflects back to space, is generally about half that of a surface of desert or dry grass. Bearing in mind that the earth is warm in the first place because it absorbs solar energy, increasing that absorption by growing forest would tend to warm it even more.

For one reason or another this is not a phenomenon that has been examined in detail by climate scientists, but rough back-of-the-envelope calculations suggest the following. In order to grow enough trees to negate the human input of extra CO_2 over the next generation, it would be necessary to increase the total area of forest by an amount roughly equivalent to two-thirds the area of the United States. The impact on global albedo would depend on exactly where in the world the trees were grown, and in particular on the albedo of the land that was being forested. However with some assumptions, it is easy enough to calculate that the overall decrease in global albedo would tend to raise the world's temperature by something of the order of a goodly fraction of a degree. The 'order of' depends a lot on what one believes might be the truth about the feedbacks of the system as discussed in an earlier chapter. Whatever, it seems highly likely that the increasing darkness of the planet could wipe out a lot of the benefit of reducing the concentration of atmospheric CO_2.

Fertilizing oceans is potentially a much easier and more effective means for disposing of CO_2. The idea is not new and was first suggested about 20 years ago by John Martin in the United States. He envisaged sprinkling iron sulphate in specially chosen ocean waters so as to trigger very large plankton blooms. The billions of cells produced in the process could absorb enough carbon from the atmosphere to cool the earth. With tongue in cheek he went so far as re-coin a famous quote in a new form - "Give me half a tanker of iron, and I'll give you an ice age". Since then there have been a number of experimental trials in various oceans, and all of them more-or-less confirm that the original idea was not stupid. Any uncertainty about the science of the process is generally related to the as yet unanswered question as to whether the cells produced in the bloom ultimately fall into waters deep enough to ensure that their carbon is not quickly returned to the atmosphere.

Several commercial companies have been formed to exploit the concept. They hope to use the technique as a mechanism to lay hands on some of the 'carbon credit' money that is likely to appear in a new era of carbon trading. How they might establish to the powers-that-be just how much carbon they had sequestered in the ocean is an interesting question. As well, they face the much more problematic task of convincing politicians that active fertilization will not unbalance the whole ecology of the world's oceans. Such an outcome is extremely unlikely, but the environmental movement has managed to create a belief within the general public that any and all algal blooms are highly toxic. This despite the fact that artificially increased concentrations of iron within the chosen areas of ocean need be no greater than the natural concentrations of iron-rich waters downstream of winds carrying dust from the continents. Such waters are generally highly productive of marine life.

Even research on the subject is an absolute no-no to the hard-core environmentalist. The Ecological Internet lobby group for instance implored us recently in an Action Alert Update to "stop rogue German ship from fertilizing the Southern Ocean in dangerous geo-engineering experiment". The ship concerned was the RV Polarstern, a German government research vessel apparently intent on "constructing a

'Frankensphere' with dramatic unknown consequences rather than reducing carbon emissions". The reaction of the environmental movement is enlightening in a way. It argues a mind-set that is not really concerned with doing something about climate change as such. Any old solution to the problem will not do. It has to be a solution involving limitation of the use of cheap fossil fuel.

Iron fertilization of the ocean would probably be the cheapest of the CO_2 disposal options if it worked as hoped by its proponents. It would certainly be immensely cheaper than simply burying the stuff. It should as well have benefits in terms of ocean productivity. Unfortunately the whole business is shaping up as an ever more polarized theoretical argument between proponents and opponents. Ultimately no doubt it will be gridlocked over philosophical matters that have little to do with the real issue. It will become a playground for lawyers to hone their skills with regard to the Law of the Sea.

On Population Growth

While there are all sorts of caveats in particular circumstances, there can be little argument with the very general proposition that the total global anthropogenic emission of CO_2 is proportional to the product of the total human population and its average standard of living. One can perhaps argue that increased standard of living need not necessarily be related to fossil fuel consumption in the future, but it is hard to deny that, other things being equal, more people necessarily implies a requirement for more CO_2 emission. And it is also hard to deny that there are a fair number of other shorter-term environmental and sociological problems which, again with other things being equal, are an increasing function of the number of people who create them.

So there is at least a superficial case to be made for considering active measures to encourage smaller population growth – this as part of an overall strategy for limiting global warming. Whether there is any real validity to the case depends among other things on deep religious and ethical considerations which are way outside the purview of this book.

Some "Model Outputs"

Bearing in mind the massive assumptions that must be made in order to produce actual numbers from the sort of hybrid climate-economic model envisaged here, the present author is a little ashamed to admit he was involved about 20 years ago in an attempt to do just that. The attempt used information gathered at a three-day workshop held in Virginia in the United States. The workshop participants were more-or-less held to ransom. They were given a broad scenario of global warming and told that, despite their protestations to the effect that the task was impossible, they had to produce their best guesses as to the costs and benefits to a number of individual nations of various techniques of global-warming mitigation.

We won't go into embarrassing detail of the model and its assumptions. A few very broad conclusions were drawn from the exercise, and most of them were obvious enough in hindsight. They did not really require development of a complicated numerical model.

One of the conclusions concerned conversion of coal-burning power stations to, say, nuclear power stations. If it were done in a hurry, within a decade or two for instance, it would be by far the most expensive option for reducing carbon emissions. The cost per person in countries like the USA and Australia of reducing the national emission by 5% would be something of the order of a thousand dollars. The operative phrase here is 'in a hurry'. The costs would drop dramatically if the changeover were allowed to occur at the end of the natural lifetime of existing infrastructure.

A broader conclusion was that, from a strictly economic point of view, it is extremely difficult to justify a costly preventative action taken purely on the basis of limiting climate change. This is particularly true for developed nations where agriculture – the main climate-sensitive activity of mankind – probably accounts for less than 10 percent of gross national product. For such nations, reducing CO_2 emission involves tampering with energy-based economic sectors which, taken as a whole, generally account for a very much larger fraction of gross national product. In any event it is true for just about every nation simply because the 'economic' rates of discount for the future are of the order of five to ten percent per year.

Perhaps the most interesting general outcome from the modeling machinations – and again it is obvious enough in hindsight when one thinks enough about it – is that the rate of discount for the future is by far the most important factor determining whether the long-term benefit of any preventative action is greater than the cost. Give or take a bit, and more-or-less whatever the cost of the action, a discount rate less than about one percent suggests that present action for future climatic benefit might indeed be worth taking. Provided of course that it is certain that the climate change would be detrimental. A discount rate of more than one percent effectively ensures that there would be no sense to present action if it were taken purely for the purpose of reducing global warming. It is not surprising that the activists' battle to limit climate change relies heavily on a concept that our current environment is the best of all possible worlds and must be preserved into the very far distant future.

Let us be more direct about this. Whether or not society should do something about global warming boils down to whether it can be persuaded of two things. First, it must be persuaded that the coming of global warming is certain and that it will be detrimental. (Actually we have two separate things right there, but let us not quibble). The scientific community has relegated that particular task of persuasion to itself. Second, society must be persuaded of the greatness of the moral virtue attached to active personal sacrifice for the benefit of people more than 4 or 5 generations into the future. That task lies squarely in the province of environmentalists and politicians. However, as a backstop in case they should fail, scientists are working hard to find arguments which suggest climatic disaster within a few decades rather than a few centuries. Bearing in mind all the uncertainties inherently associated with climate research, and recognizing that in such circumstances it is almost impossible to prove anything right or wrong, no doubt it will be easy enough to find arguments that fit the required bill.

Inevitably we come back to the point that the whole case for doing something about climate change is based almost entirely on the need to maintain the quality of the environment. There are of course very good arguments for so doing. The problem emerges when the need is transformed into one of maintaining the environment exactly as it is. Change of itself is not necessarily a bad thing. On past experience, it is probably essential for the continued existence of humankind.

Unfortunately the argument has indeed descended into one of preserving the status quo. This in turn has led to a propaganda ploy that goes down well in most societies because it implies a sort of uniqueness - namely, that one's own country is peculiarly sensitive to climate change and will suffer most in the coming Armageddon. The origins of myths like these are almost impossible to identify. They just seem to appear out of nowhere.

In the Australian context one presumes such a story can be sold easily enough because the continent has lots of desert, the growth of plants is generally water limited, and the models of climate (or at least the models of climate we hear about) produce scenarios of much dryer times in the future. We don't hear about the fact that over the last few decades Australia has in fact got greener. Or that increasing CO_2 in the atmosphere should lead to greater efficiency of plants in their use of water.

A Couple of Bottom Lines

In the early days of the global warming business there was a reasonably sensible attitude around the traps to the effect that, in view of the uncertainty of the forecasts of disaster, any expensive preventative action should be undertaken only if it were worth doing for other shorter-term reasons as well. Improving the efficiency of transport would seem to fit that criterion, as perhaps would an eventual move away from electric power stations fired by fossil fuel.

For some strange political reason, the attitude has been lost in the modern frenetic rush to do something immediately – that is to say, as Bertie Wooster would have it, "eftsoons and right speedily". We are seeing a large fraction of an already vast expenditure on climate being put towards research concerned with carbon sequestration. With burying the stuff, that is. It is hard to imagine how we will be able to put piles of buried carbon to good use for other purposes. It seems there is indeed an innate preference by humankind for options requiring the maximum of self-flagellation.

On another issue, it is worth referring again to the work of the Professor Garnaut.

His calculations involved running models of Australian climate one hundred years into the future. He used the forecasts so obtained to drive other models of the national economy over the same period. One is irresistibly reminded of the blind leading the blind. Never mind. Setting that aside, the main result of all the activity was the calculation of a negative impact of reduced rainfall on important agricultural regions such as the Murray-Darling basin. It seems that this will cause a 10 percent reduction (below what it would otherwise have been) of the material welfare of the average Australian. As a consequence we must 'act now' by running around like a lot of demented souls telling everyone that the end of the world is nigh. And we must turn the economy upside down at the same time.

Well, maybe. It all appears terribly naive. Particularly when in the same breath Professor Garnaut tells us that, even when all the disaster of climate change has come upon them, the Australians of the early 22nd century will be nearly four times wealthier than we are.

CHAPTER 4

SOME RANDOM SOCIOLOGY

"...no science is immune to the infection of politics and the corruption of power.... The time has come to consider how we might bring about a separation, as complete as possible, between Science and Government in all countries. I call this the disestablishment of science, in the same sense in which the churches have been disestablished and have become independent of the state."

Jacob Bronowski (1908-1974)

Perhaps the most interesting question in all this business is how it can be that the scientific community has become so over-the-top in support of its own propaganda about the seriousness and certainty of upcoming drastic climate change. Scientists after all are supposed to be unbiased in their assessment of a problem and are expected to tell it as it is. Over the centuries they have built up the capital of their reputation on just that supposition. And for the last couple of decades they have put that capital very publicly on the line in support of a cause which, to say the least, is overhung by an enormous amount of doubt. So how is it that the rest of the scientific community, uncomfortable as it is with both the science of global warming and the way its politics is played, continues to let the reputation of science in general be put at considerable risk because of the way the dangers of climate change are being vastly oversold?

To set the scene and to make the point that the overall issue is not simply an esoteric argument of no practical consequence, it is perhaps worth relating a few incidents from personal experience of the game. None of them is in any way unique. All of them could be topped easily in a 'good story' competition open to any other scientist who is vaguely sceptical about global warming hysteria. All of them are probably worth telling only in the context of scientific research, since that activity is particularly reliant on its reputation for immunity to the forces of the politically correct. In the long term it is particularly reliant also on the very existence of sceptics both within and without its ranks.

The first is an example of the perennial story about restriction of public comment by staff other than official public relations personnel. Such stories can be found in virtually any organization. In most cases the restriction is both justifiable and sensible. In the case of scientific research, and particularly in the case of climate research that falls squarely inside the category of 'public good' science and is performed mainly by government laboratories, it is not nearly so justifiable and is certainly not sensible. Unfortunately, laboratories concerned with the climate issue have to be extremely careful about relations with their source of funding since there is normally no other source of any significance. In the same manner as their financial master, they need to be careful not to step too far away from alignment with public belief. So it is not unusual for climate scientists to mumble darkly - and occasionally even publicly after they have retired - that their organization is not backward in stifling comment from scientists on matters that bear on its political aspirations.

It was ever thus. In the early nineties I was involved in setting up an Antarctic research centre, which was, and still is in a slightly different form, a sizeable research institution specifically designed to examine the role of Antarctica and the Southern Ocean in global climate. I made the error at the time of mentioning in a media interview - reported extensively in 'The Australian' on a slow Easter Sunday - that there were still lots of doubts about the disaster potential of global warming. Suffice it to say that within a couple of days it was made very clear to me from the highest levels of CSIRO that, should I make such public comments again, then it would pull out of the process of forming the new Centre. Since CSIRO was a major partner in the venture and its withdrawal would have killed the whole thing dead, it is perhaps not surprising that I found the message to be both cogent and effective. It turned out that the response was related to efforts by CSIRO at the time to abstract many millions of new climate research money from the federal government.

The story illustrates one of the reasons why it is no accident that much of the limited amount of publicly expressed scepticism about the disastrous nature of climate change comes from scientific retirees.

As for instance from Dr. Brian Tucker, a former Chief of CSIRO's Division of Atmospheric Research who was a specialist in numerical climate modeling and therefore knew better than most where the bodies are buried in the climate change game. He kept remarkably quiet about his worries on the matter. Then he retired, and for four or five years thereafter was the bane of the global warming establishment because of his very public stance against many of its scientific sacred cows. Eventually the system managed to limit the damage by labeling him as one of the usual suspects who was now out-of-date and in any event was probably on the payroll of industry. It should be noted that 'usual suspect' is not nowadays just a term of slightly exasperated endearment. In the context of climate change, it is a label carefully chosen to sideline those so lacking in virtue as to question the influence of the politically correct over the scientifically naïve.

The lesson to be learned from all this is that politicians face something of a problem in their search for unbiased scientific advice on matters to do with climate change. In Australia for instance, if CSIRO and the Bureau of Meteorology combine to present a united front on global warming (as they have formally done in the last year or two by creating a joint climate research establishment) there is nowhere to turn for a second opinion other than to a few scattered retirees of no great account. Because the government is not entirely made up of idiots, it must know perfectly well that its departments and agencies can be as subject to bias as any of the relevant industry associations. As always, livelihoods are at stake. But there isn't a lot that can be done. For the sake of good order, governments are more-or-less obliged to accept without serious question the advice of their own agencies.

The situation in the US is slightly better in that there are climate research groups outside the direct funding umbrella of either government or industry. Certainly there are a number of well-funded think tanks which can lay their hands on scientific advice on climate change which runs counter to bureaucratic policy, and on scientists who can provide effective independent advice to influential congressmen who may wish to get involved in the debate.

Which leads to a more modern tale, told here mainly because it bears specifically on the difficulties faced by politicians in their search for unbiased advice on climate change. It is an example of the strength of the forces bearing on advisory bodies – even on independent advisory bodies – to conform to the establishment view. In the case of scientific advisory bodies, the forces are amplified to some extent by the general desire of scientists to be gentlemanly, politically responsible, and visibly democratic. We could argue perhaps that none of these characteristics is essential, or even desirable, in scientific research, but we won't bother. The tale is also an indication that in Australia at least, because of its relatively small population, there is a considerable degree of inbreeding among scientific advisory bodies and the relevant agencies of government.

In 2008 when Professor Garnaut was producing his draft and final reports on what should be done about climate change, an Academy of Science ad-hoc committee was charged with the task of producing an 'Academy response' to the Garnaut reports. The Academy is roughly the equivalent of the Royal Society of Great Britain or the Academy of Sciences in the US, and consists of several hundred 'Fellows' drawn from the elite of all the active scientific disciplines in Australia. Its pronouncements carry considerable weight in the corridors of power. While it receives some funding from the federal purse, it is supposedly an independent body not beholden to the government or to any of its departments and agencies. Its advice should therefore be reasonably independent and unbiased, which is presumably why it is valued by society, and why, in fact, funding from the federal purse is made available to it.

The committee of six had some small interaction by e-mail over the several weeks prior to its single meeting in Canberra. There was some discussion with Academy officers early on the day of the meeting as to what purpose an Academy response might serve, bearing in mind that the major climate research agencies were perfectly capable of selling their own message on the matter both to the public and to the politicians. It was agreed, although without obvious enthusiasm by the officers, that the purpose was to highlight any aspects of the Garnaut reports which, in the view of the committee, were inadequate in their coverage, emphasis or veracity.

The committee spent the day in the frenetic rush usual in such circumstances. Nevertheless it produced a rough draft of a response that indeed made a number of points worth putting into coherent English for distribution as a formal document. Among them were the usual pleas for more money to do research on this or that subject dear to the heart of particular members of the committee. Among them also were one or two significant issues of broader concern. Perhaps the most important was a deliberately worded statement about the committee's worries with the accuracy of the forecasts of Australian regional rainfall one hundred years into the future. These forecasts after all were the major input to the Garnaut calculations of the negative impact of global warming, and therefore to the Garnaut recommendations about how the nation should respond.

While the committee was aware of all the 'ifs' and 'buts' of 100-year prediction of rainfall, it was aware too of the delicacy of saying so in an Academy response. But if indeed there is something of the order of a 50/50 chance that the forecasts supplied to Garnaut were nonsense, then it seems reasonable that the fact should be made known in plain English, and by a respectable authority, both to the political decision makers and to the public at large.

The day ended with a recommendation by the Academy officers that the draft response be discussed with Professor Garnaut himself before proceeding. This seemed a trifle peculiar to one or two of the more simple-minded of the committee, but the technique is a standard tactic to ensure that matters of potential controversy are headed off at the pass. In any event it was arranged for the Academy officers to meet with Professor Garnaut a few days later.

Rumour has it that sometime during the meeting Professor Garnaut became very sympathetic to the need for vast new resources to address the need for basic research in the field of climate science. From that point on, the discussion was all sweetness and light. And in the following weeks there was considerable discussion within the corridors of scientific power about how to present the conclusions of the committee. In the end it seems that the idea of a response to the Garnaut report was dropped altogether.

Now to be fair, one should not read too much into the outcome. No doubt there were a number of logistic and timing issues that affected the decision. And in fact the Academy eventually did come out with a formal statement of priorities for Australian climate research which contained brief reference to the fact that Professor Garnaut had used CSIRO-BoM rainfall projections and that these projections were fairly problematic. There was accompanying careful comment that "economic modeling based on one scenario that does not consider the recognized uncertainty of climate models therefore has the risk of reaching incorrect conclusions".

True enough of course, but not exactly the sort of comment that grabs attention. It is a pity that the Academy lost the opportunity to make a much more forceful, and equally true, statement to the effect that Professor Garnaut might just as well have tossed a penny labeled "much dryer" on one side and "much wetter" on the other, and selected the first toss that gave him the forecast he wanted.

The system, it seems, is designed to look after its own. Even the Academy has to watch its step in the mire of political expediency.

Going off at a complete tangent for a moment, the Academy statement on priorities talks among other things about the need for more money to restore the Australian climate monitoring network, which is indeed something that sorely needs to be done. It talks about the need for more basic research, and in particular recommends huge new computing resources so as to upgrade the present climate-modeling capability. This last is a standard, and highly arguable, recommendation which emerges all over the world from just about any official consideration of the needs of climate research. The trouble is that it has yet to be established exactly what is the scientific problem which, if solved, would lead to more believable, or indeed just plain believable, long-range climate forecasts. Until that problem is precisely defined, it seems a little strange to be demanding computing resources vastly bigger and better than those currently available. They would simply ensure continuation and expansion of current activity in the vague hope that 'more of the same' will provide a scientific breakthrough some time in the future.

A third very minor incident is an example of the sort of thing most scientists have come across at one time or another – namely, off-the-cuff statements by quite senior scientific movers and shakers who simply cannot hear what they are saying in the context of what scientific research is all about. They cannot hear their own wielding of the bat of political correctness.

This particular incident occurred at a small conference in South Australia at just about the time Professor Garnaut had been given the task of advising the newly elected Australian federal government on what to do about climate change. A senior scientist of my acquaintance spent most of his time at the conference huddled over his computer e-mailing back and forth and correcting versions of a document that was to be the climate agency input to the Garnaut enquiry. He was obviously furiously busy. When I asked what it was all about, he maintained during his explanation that the greatest problem was to ensure that whatever was said in the document gave no comfort at all to 'the vast army of sceptics out there' who could not be made to believe in the seriousness of the global warming story.

His comment is an indication of a couple of the basic truths behind the climate-change debate. The first is that the scientists pushing the seriousness of global warming are perfectly well aware of the great uncertainty attached to their cause. The difficulty for them is to ensure that the lip service paid to uncertainty is enough to convince governments of the need to continue research funding, but is not enough to cast real doubt on the case for action. The paths of public comment and official advice on the matter have to be trodden very carefully. The second basic truth is that there is a belief among scientific 'global warmers' that they are an under-funded minority among a sea of wicked sceptics who are extensively funded by industry and are closely related to Satan. The difficulty for them is to maintain a belief in their own minority status while insisting in public that the sceptics, at least among the ranks of the scientifically literate, are very few.

Now scepticism may have its faults, and may indeed be inappropriate in all sorts of circumstances. On the other hand it is the very lifeblood of science, and statements about giving no comfort to sceptics are almost funny when come across in a research context. I really thought my acquaintance was joking for a while until it became obvious from his later conversation that he was actually quite serious.

The attitude is unfortunately very common in the scientific establishment. Perhaps one of the worst (best?) examples concerns the behaviour of the Royal Society in Great Britain as reported extensively in the Guardian newspaper a couple of years ago just before the issuance of the fourth IPCC report. The Society requested in a formal letter to Exxon-Mobil that it cease funding organizations that have "misrepresented the science of climate change by outright denial of the evidence". A reasonable enough request at first glance one might think. Well, sort of, except that, as one of the most senior of the world's scientific societies, it should know better than anyone that the evidence for an upcoming disastrous impact of man-induced global warming is far from solid. The actual numbers that are quoted about the seriousness of global warming derive purely from theory. Theory is evidence of a sort, but it is pretty ropey stuff until it is experimentally verified.

It is therefore very peculiar that the Society should attempt to stifle sources of funding for organizations that dare to question what necessarily has to be a doubtful, and potentially an enormously expensive, prediction of the future. One is almost bound to believe in a hidden agenda.

The Royal Society letter took on a more sinister aspect when later it said: "To be still producing information that misleads people about climate change is unhelpful. The next IPCC report should give the people the final push they need to take action and we can't have people trying to undermine it". The staggering thing is that the Society, which in other circumstances would be the first to defend the cause of free enquiry, and in the same fashion as the scientist at the conference in South Australia, seemed not to be able to hear what it was saying.

Not allowing people to undermine the call to action inspired by an upcoming IPCC report is most revealing. It illustrates something that has been commented upon by others – namely, that the scientific statements on global warming coming out of Britain (and nowadays out of many other countries as well) are apparently a reflection of government policy rather than the other way around.

Finally on the subject of the deafness of scientists to what they are saying, on many occasions I have heard the chairmen of climate-related conferences publicly express their relief at the absence of the press. It seems that any good reporter would "seize upon the controversy and uncertainty that has emerged from our scientific discussion and report it in such a way as to undermine public belief in the seriousness of the global warming problem". Reporters, it would appear, are almost as wicked as sceptics.

A fourth story is a small example of the difficulty faced by scientists these days if they try to publish research results that are not supportive of the established view of greenhouse disaster. It can be bettered by any number of similar tales from climate scientists who have tried to get non-supportive results published in respectable research journals. To be fair, it is not an unusual story in any scientific discipline. 'Peer review' is far from an ideal process, and generates an infinite variety of complaints from disappointed authors whose manuscripts are rejected by the hard-nosed editors of research journals.

Three of us sent off a manuscript in early 2008 to a climate journal. It was a simple-enough article reporting the existence of long-term trends of humidity at various heights in the atmosphere as they (the trends) appear in a well-known and much-used set of data concerning atmospheric behaviour over the last five decades. The data are readily available on the web and have become something of a workhorse for much modern research on meteorology and climate.

The manuscript did two things.

First, it pointed out that, according to the data if they are taken at face value, the humidity over much of the world has decreased over the last several decades at altitudes above about 3 kilometres. The humidity information derives ultimately from balloon-borne instruments called radio-sondes which are launched every day from the international network of meteorological stations. Some of the stations have been operating for 50 years.

One must say immediately that radiosonde humidity measurements have more than their fair share of problems. They are not exactly the most accurate measurements ever made.

Second, it made the point already discussed in an earlier chapter that water vapour feedback in the global warming story is largely determined by the water vapour at heights above 3 kilometres. Total water vapour of the atmosphere may increase as the temperature of the earth's surface rises, but if at the same time the upper-level humidity decreases then water vapour feedback would be negative. If, repeat if, one could believe the radiosonde data 'as is', water vapour feedback over the last few decades has been negative. And if the pattern were to continue into the future, one would expect overall water vapour feedback in the climate system to halve rather than double the temperature rise due to increasing CO_2.

Our article concluded by suggesting that, in view of the significance of upper-level humidity to the climate change story, and despite all the potential problems with the raw information, the international balloon data on upper-level humidity need to be subjected to detailed re-examination along the lines of that presently going on for the surface temperature-measuring network. (That network also has lots of problems of accuracy, and there is a major international effort devoted to abstraction of real trends from highly suspect data. Despite popular belief, there is still much doubt about the magnitude of the apparent rise in global temperature over the last century).

Suffice it to say that our article was knocked back. This largely because of an unbelievably vitriolic, and indeed rather hysterical, review from someone who let slip that "the only object I can see for this paper is for the authors to get something in the peer-reviewed literature which the ignorant can cite as supporting lower climate sensitivity than the standard IPCC range".

We argued a bit with the editor about why he took notice of such a review. We are not exactly novices in the research game, and can say with reasonable authority that when faced with such an emotive review the editor should simply have ignored it and sent the paper off to someone else. The argument didn't get far. In particular we couldn't get a guarantee that a re-submission would not involve the same reviewer. And in any event the conditions for re-submission effectively amounted to a requirement that we first prove the models and the satellites wrong.

A couple of weeks after the knock-back, and for unrelated reasons, two of us were invited to a small workshop on atmospheric humidity held at Lamont Doherty Earth Observatory in New Jersey, whereat we told the tale. The audience was split as to whether the existence of the radio-sonde trends in humidity should be reported in the literature. Those 'against' maintained loudly that the data were too 'iffy' to report the trends publicly in a political climate where there are horrible people who might make sinful use of them. Those 'for' could only say that scientific reportage shouldn't be constrained by the politically correct – particularly when there are no other sources of experimental information on the subject except for rather problematic satellite measurements. The matter was dropped. We found after the event that the journal editor had come (we think specifically) to hear the talk. He didn't bother to introduce himself.

The story itself doesn't amount to much. It is significant only as an example of the fact that, because virtually all measured long-term trends in climate have problems with regard to accuracy and believability, there is always the opportunity to write them off if one is so inclined. This is where the population density of the 'so inclined' is a significant factor in whether or not a particular result gets accepted by the general scientific community.

Perhaps the story is significant also in that it shows how naïve we were to imagine that climate scientists would welcome the challenge to examine properly and in detail even the smell of a possibility that global warming might not be as bad as it is made out to be. Silly us.

A Little History

Research on climate became a formal international programme in the early seventies. It was, and still is, based on a network of committees to whose meetings the climate scientists of the world flock like homing pigeons every northern summer. It grew out of an earlier, highly successful, programme of the World Meteorological Organization designed to establish how far ahead one might be able to make detailed predictions of the weather. The answer to that particular question turned out to be about 10 days.

The new programme lacked some of the characteristics of the old. It lacked focus on a specific problem, so progress was hard to define and to measure. It lacked an experimental centrepiece to which people could point with some pride and say that things were indeed moving. And to some degree it lacked the powerful personalities who had the clout to keep things on the rails by virtue of their status within their home research organizations. Scientists, while maintaining their public disdain of activity that kept them away from their 'real work', just loved it. The vast array of international meetings ensured that they could be numbered among the most travelled people on Earth.

It soon became apparent that the problem of short-term climate forecasting - that is to say, predicting what will be the average sort of weather over the next year or two - might very well be inherently insoluble. It was therefore a little dangerous to build an expensive and well-publicized international research programme about it. On the other hand it did seem that it might be possible to forecast the very broadest aspects of climate many decades and centuries ahead. Intuition that this was so had enabled scientists a hundred years earlier to flag the possibility that global warming would result from increasing concentrations of atmospheric carbon dioxide. It was this very long-term issue that finally emerged as the main public face of what was by then the World Climate Programme of the United Nations.

The predictions up to that time (this was the early eighties) were of a rise in temperature of a degree or so over the next one or two hundred years. Such numbers didn't sound too horrific to the man in the street. He was a fairly practical chap with bigger problems on his mind. So during the eighties and early nineties there began an almost subconscious search for reasons why the rise in global temperature might be much greater than a degree, and why the change might occur over less than 100 years. The search was strongly linked to the rapid development of computers and computer modelling, which in principle provided for the first time the tools to simulate the vast number of complicated processes that determine Earth's climate.

As discussed in an earlier chapter, it needs to be understood that any reasonable simulation even of present climate requires computer models to be tuned. They contain parameters (i.e. pieces of input information) whose numerical values are selected primarily to give the right answer about today's climate rather than because of some actual measurement of their value. This was particularly so in the mid-eighties. The problem with tuning is that it makes any prediction of conditions different from those of the present far less believable. Even today the disease of 'tuneable parameters' is still rampant in climate models, although fairly well hidden and not much spoken of in polite society. The scientifically-inclined reader might try sometime asking a climate researcher just how many such parameters there are in his or her latest model. He or she (the reader) will find there are apparently lots of reasons why such a question is ridiculous, or if not ridiculous then irrelevant, and if not irrelevant then unimportant. Certainly the enquirer will come away having been made to feel quite foolish and inadequate.

Different modellers developed their own particular ways of simulating the processes determining climate, and used as well their own values of tuneable parameters. Thus emerged a highly satisfying spread of the forecasts of the likely rise of Earth's temperature over the next hundred years. Some were as high as six or seven degrees, which is getting on for being almost as large as the rise in temperature when the world came out of the last ice age. And while other scientists might have had an intuitive feeling that such forecasts were way over the top, there was (and for that matter still is) no way to prove unequivocally that such forecasts are wrong. The complexity of the models is so great that it is extremely difficult, even for other climate modellers, to establish exactly why one model should give a vastly different answer to another – let alone establish which is the more likely to be correct.

So the habit arose of quoting the range of the different forecasts without any discussion of the merits of one versus another. Some researchers were uncomfortable about it, but soothed their consciences by insisting that the various results were only 'scenarios for scientific study' rather than actual forecasts of what might be expected to happen. The insistence was more than a little naïve (or perhaps more than a little cunning) since the distinction went quite over the head of the general public. It was of course the extreme scenarios which got into the folklore of public opinion.

At the same time there began a fair amount of research on the sensitivity to climate-change of this or that biological ecosystem and of this or that human activity. Again the uncertainty of the science became something of a benefit when trying to build a case for action. "So and so's research indicates that even a small rise in temperature might be disastrous to the yellow-bellied sap-sucker of the upper Zambesi. It is true that the research concerned only one sap-sucker which died in an artificial hothouse sometime after being slightly chewed by a cat, and more work needs to be done, but can we afford to ignore the result?" Or words to that effect. Suffice it to say that there arose enough semi-scientific myth and legend about the possible detrimental effects of small climate changes that it was easy to contemplate doomsday in the face of somebody's mooted six or seven degree rise in global temperature.

The old adage that bad data are worse than no data was largely forgotten.

The IPCC

So there was generated enough noise about the disastrous effects of climate change to justify the establishment in the late eighties of the IPCC - the Inter-governmental Panel on Climate Change. The organizers set up a report-producing mechanism which involved three separate international working groups. The first dealt with the science behind the predictions of global warming, the second concerned the potential impacts of the warming on society, and the third examined the various possibilities for political and social response.

Right from the outset it was clear that only the third group was really significant, which was why the US made such a fuss at the time to ensure that its first chairman was an American. They realized that the whole IPCC process could very easily be turned into a cudgel with which to beat the west in general and the USA in particular. I well remember an American colleague telling me in 1990 that the developing countries were quite up-front about the matter. "You westerners" he reported them saying at one of the early IPCC meetings "have been exploiting us for hundreds of years, and here finally is an opportunity to redress the balance. Give us all your money!" One assumes he was laying it on a bit thick, but recent history and the developments associated with the Kyoto Protocol suggest that there was more than a modicum of truth to his view of things.

The main reason for the existence of the first two working groups was - and still is - to lend gravitas and respectability to the essentially political deliberations of the third.

Re-vamped IPCC scientific reports have appeared with much fanfare every few years since 1991. Each of them is accompanied by a "Summary for Policy Makers" which is the only thing read by 99.9%

of those concerned with the matter. It is the real determinant of public and political opinion on the greenhouse issue. It is also that part of the report whose wording is more-or-less beyond the direct influence of the average scientist.

Each of the successive summaries has been phrased in such a way as to appear a little more certain than the last that greenhouse warming is a potential disaster for mankind. The increasing verbal certainty does not derive from any particular advance of the science. Rather, it is a function of how strongly a statement about global warming can be put without inviting a significant backlash from the general scientific community. Over the years, the opinion of that community has been manipulated into more-or-less passive support by a deliberate campaign to isolate - and indeed to denigrate - the scientific sceptics outside the central activity of the IPCC. The audience has been actively conditioned into being receptive. It has thereby become gradually easier to sell the proposition of greenhouse disaster.

The conditioning process plays largely on the belief that, because the IPCC is an international body responsible to lots of national governments, and because it is serviced and constrained by many supposedly apolitical scientists, it is somehow free from the biases either of industry or of political correctness. That belief (at least with regard to political correctness) is quite naive. Most of the developed countries have institutionalised their greenhouse activity within government agencies devoted specifically to mitigation of global warming. Their budgets are enormous. It is not likely that the public servants who staff them will be receptive to doubts about their reason for existence. Nor, for that matter, are the actual research institutions concerned with global warming likely to bite the hands that nowadays feed them.

The conditioning process also involves spreading the belief that IPCC announcements are the consensus opinion of the vast majority of knowledgeable climate scientists. The belief is simply not true - at least not in the sense that the public understands it.

It is true enough that there are a fair number of scientists involved with the IPCC reporting process. This is a necessity because the study of climate involves a great number of specialized and mostly unrelated disciplines, and each discipline needs to be represented. On the other hand there are very few overall experts on the subject. It is far too broad. So the consensus is not really about the science of the matter, because the individual specialist knows nothing (or nothing much) about most of it. Rather, and to the extent that consensus exists at all, it is simply a public expression of faith in the profession. It is a public expression of the hope that there is somebody else within the system who is knowledgeable enough to pick up any major errors of fact or philosophy.

And in any event, climate scientists are far too diffuse a constituency to mount a coherent disagreement. Whatever their private worries about the overall conclusions, they won't publicly disagree with them provided the wording stays within the bounds of the pre-conditioning.

Some General Worries

The increasing shrillness of the message about global warming has about it a certain messianic flavour usually associated with religious faith rather than empirical or scientific knowledge. And when one looks below the surface of the science, it is certainly true that a rather large amount of faith is required to go along with the doomsday predictions which have been sold to the public.

First let it be said again that even 'back-of-the-envelope' calculations (much more believable in many ways than large computer models) suggest that there will indeed be some degree of global warming because of increasing greenhouse gases in the atmosphere. So there is not much argument from scientists about the actual existence of the phenomenon. On the other hand there is very great scepticism - this despite contentions to the contrary by the IPCC and the institutions related to it - that the amount of the warming will be enough to worry about, or indeed enough to notice, bearing in mind the natural variability of both climate itself and the ecosystems on which it bears.

What worries most scientists who know a little about the subject is that virtually the entire experimental support for the theory of global warming is based on the rough co-incidence of the slight rise in Earth's temperature over the last 100 years with the rise in concentration of atmospheric carbon dioxide during the same period. The coincidence needs to be viewed in light of the fact that the world's temperature has always in the past gone up and down like a yo-yo on all sorts of time scales. Presumably it will continue to do so into the future.

This is why the 'hockey stick' reconstruction of past temperatures became so famous so fast. It seemed to show that temperature in fact didn't change much for 900 years, so the rise over the last hundred years must be something really quite out of the ordinary. Since this is what scientists wanted to hear, the statisticians who pointed out the problems with the data had to go to enormous lengths to report their results and to cover their backs. The vengeful wrath of the greenhouse community can be quite dreadful. And indeed the white blood cells of the IPCC are still gathering, and it is a moot point whether the problems with the hockey stick will ever be admitted. One suspects that, at best, they will simply be ignored. Around the traps one hears comments to the effect that "Oh yes - the original authors of the hockey stick business were probably a bit sloppy with their data, but it doesn't really matter to the overall greenhouse issue." Perhaps it shouldn't matter, except that the IPCC made such a big deal of the results at the time.

The case for climate change is not being made any more acceptable in the eyes of the non-committed general scientist when virtually any long-term change of an ecosystem is attributed more-or-less automatically to global warming. It is the easiest way to 'explain' an observation of (say) a decline in population of a species, and has the additional advantage of being more likely to grab the headlines. It is certainly much easier to blame climate change than to put the observation into the context of an ecosystem which has evolved to be naturally variable - turbulent if you will - and in which the populations of individual species fluctuate naturally as a survival mechanism.

What worries most scientists who know perhaps rather more than a little about climate is that, even after at least 30 years of hard work, the numerical models tell a coherent story about the possible change of climate in only one respect - namely, that the global average surface temperature is likely to go up. But we knew that more than a hundred years ago. As discussed in Chapter 2, when it comes to prediction of more detailed information like the *distribution* of temperature and (particularly) of rainfall, then individual models tell entirely different stories. One model might suggest that Australia, for instance, will become dryer. Another might suggest that it will become wetter. A third will calculate no change. Needless to say, in a country not exactly over-endowed with water, it is the 'dryer Australia' scenario which hits the headlines or makes it into the scientific literature, even though actual measurements over the last hundred years seem to indicate otherwise.

The bottom line is that most scientists - certainly most physicists and mathematicians who are more used to such things - strongly distrust large-scale numerical models which rely heavily on tuneable parameters and other artificial constraints to keep them from going haywire. Particularly do they distrust the output of models whose predictions cannot, for lack of sufficiently detailed data, be tested against real events of the past.

So one suspects that a fair amount of the shrillness of the climate message derives from a fear that something will happen to prick the scientific balloon so carefully inflated and overstretched over the last few decades. But the IPCC doesn't really need to worry. The difficulty for the sceptics is that credible argument against accepted wisdom requires, as did the development of the accepted wisdom itself, large-scale resources which can only be supplied by the research institutions. Without those resources, the sceptic is only an amateur who can quite easily be confined in outer darkness.

Indeed, when looked at most broadly, it is this need for massive resources before engaging in argument about the good or bad of climate research that takes the whole business outside the realm of acceptable science. It means in practice that the checks and balances usually associated with science simply can't operate.

Idealism Will Out

The typical scientist is perhaps rather more idealistic than most people. His trouble is that his profession rarely provides outlet either for idealism or for its associated activity of political activism. Research after all is supposed to be disinterested. The climate-change issue, the daddy of all environmental problems, seems to have provided an opportunity for researchers both to have their cake and to eat it too.

On the one hand, the study of climate is inherently worthwhile - this in the strictly selfish sense that the scientific problem is real (lots of research these days doesn't have that advantage), it is popular with the public, and it is relatively well funded. On the other, it provides a direct link to grand-scale social activism and to the feeling of being part of a cause. This last is extremely attractive. Most climate scientists, presumably in common with many environmentalists, are of the belief that there are vast numbers of well-funded and well-organized anti-greenhouse people out there. These are assumed to be unconscious dupes of nasty multi-national energy companies who are devoting enormous resources to manipulating public or political opinion. How good it is to be part of a small coterie of dedicated knights, each on a white horse and each armed with the lance of science, fighting heroically to save the future of the world against overwhelming forces of evil.

The balance of power is not quite as they imagine. 'The IPCC process' and the political machinations surrounding the Kyoto Protocol have spawned large government bureaucracies in most of the major industrialised nations, the staff of which are, reasonably enough, devoted to the global warming cause. The Australian Department of Climate Change is a typical example. Add the research institutions and university scientists who feed off these bureaucracies, and support them all by the environmental movement. Industry in Australia, or at least that industry which has a natural interest in the politics of climate change, may field perhaps half-a-dozen people whose part-time brief is to keep abreast of what is going on. The balance is very definitely to the side of the 'goodies'.

But it is difficult to convince a climate scientist.

CHAPTER 5

A DIVERSION INTO THE CHANGING CHARACTER OF SCIENCE

"We are raising a generation of young men who will not look at any scientific project which does not have millions of dollars invested in it ...... We are for the first time finding a scientific career well paid and attractive to a large number of our best go-getters. The trouble is that scientific work of the first quality is seldom done by the go-getters ...... The degradation of the position of the scientist as an independent worker and thinker to that of a morally irresponsible stooge in a science-factory has proceeded even more rapidly and devastatingly than I had expected."

Norbert Weiner (1894-1964)

C.P. Snow gave his 'two cultures' lecture at a time in the late 1950s which was probably the peak of the reputation and influence of science. The accelerating growth in expenditure on research was at its most obvious. The quality of the research was remarkable, driven as it was by scientists whose experiences in the second world war gave them the maturity to differentiate between significant problems and the random collection of information. Society at large subconsciously recognized the maturity and left them alone to pursue whatever it was they were doing.

In Australia certainly, the directions of the burgeoning research within CSIRO and the universities was very much driven from the bottom up. These were the halcyon days when members of the Executive of CSIRO spent a lot of their time visiting individual scientists in the laboratories, and somehow left the impression that, in the style of Yes Minister, they were 'mere cyphers who came only to service the requirements of the man-at-the-bench'. Mind you, the Executive members of the day were a cunning lot and probably consummate liars as well, but the light-touch technique was certainly appropriate and successful for its time. Meanwhile the god professors of the universities reigned more-or-less unchecked over their departments and their university administrations, and most of them made a reasonably good fist of the job. At least in the scientific departments, and presumably also in the other disciplines, there were not too many disastrous appointments to the ranks of academia.

Nostalgia isn't what it used to be of course, but things are certainly different these days. Gibbon, who was something of a bore on the subject of the fall of the Roman empire but who had the occasional nice turn of phrase, seems to have summarised the situation rather well. "A cloud of critics, of compilers, of commentators darkened the face of learning, and the decline of genius was soon followed by the corruption of taste". Managerialism, that combination of over-management, political correctness and paralysis by analysis has infected science to an incredible degree. It is particularly rampant in Australia. Scratch the typical working scientist, and his usual response will be a torrent of abuse about the management of the system in which he works.

While the plague of managerialism exists more or less everywhere and is therefore no longer all that remarkable, the interesting thing about its stranglehold on science is that in many ways the researchers have brought it on themselves. They have played the card of their own intrinsic worth rather too often, and as a consequence the public no longer trusts the judgement of scientists to the degree that it might have done in times gone by. Scientists have become too numerous, too competitive, too visible and in some cases too involved in the political process. They have, in short, become like the rest of us. The perception of their worth and character derives from the average rather than the elite, and vast systems of management have evolved to handle the lowered expectation.

At which point one should perhaps admit that we are making no attempt here to justify what may seem to be the typical jaundiced view of a member of the older generation about an activity undergoing enormous change – namely, a view that things are going to hell in a hand-basket. Rather, this chapter raises two or three issues which, with a bit of twisting here and there, bear on the widely held belief that scientific governance has become an end in itself. In turn, it bears also on the rise and rise of the climate-change bandwagon, and the difficulties faced by the average person these days in gauging the reliability of public statements by the scientific powers that be.

Too Many Scientists

In hindsight the scientific world began to change drastically in the mid-seventies when the expenditure on research became noticeable because it was by then a significant fraction of the national budget. The fraction varied a little from country to country, but in general for the developed world was something of the order of a percent or two. It has remained much the same ever since. Scientists, who of all people should recognize that exponential growth is unsustainable for very long, found it extremely difficult to recognize the new circumstance. Perhaps the most significant of their mistakes was to continue to reproduce themselves without thought for the future.

We are talking here not about scientists in general but about research scientists. Nowadays such people almost certainly have a PhD and should in principle, and with a few caveats, be able to define their own research directions within the general area of whatever is important to their employer. In the more relaxed and confident days of the middle of last century, they would all probably have imagined themselves in the 'alpha-plus' category of scientists envisaged by Snow. Anyway there was then, and still is now, an almost unquestioned belief around the traps that the training of ever-more research scientists is inherently a good thing. The inevitable result is that the country is producing far too many of them. One can argue about the precise numbers, but the overall conclusion is solid enough. On figures from earlier this decade, each year Australia is producing about 4000 PhD graduates. This is somewhere between five and ten times the number needed to fill those of its vacant positions specifically requiring a training in independent research. The annual direct cost to the federal purse of delivering these graduates is getting on for half-a-billion dollars, while the annual direct cost to the community as a whole is easily more than the full billion. And each year the output is increasing.

Now there is considerable support within the academic community for the concept that obtaining a PhD has a much wider significance than simply training for a job. The argument runs that PhD training is of value in many walks of life outside the limited market for the professional researcher. Perhaps, it is said, we should regard the PhD in much the same light as the undergraduate arts and humanities degrees of earlier times – a sort of necessary background for the truly educated person, and in turn a sort of quantitative measure of the wealth and enlightenment of the society. Surely, if there are indeed people out there who wish to take out PhDs and are capable of it, the society should provide and pay for their further education? The argument must be fairly powerful since somehow or other it has been successfully sold even to the hard-nosed rationalists within the federal government. Universities are paid twice as much for training a PhD as they are for training an undergraduate. The extra payment has been, and remains, a great incentive for universities to build the number of their higher degree candidates to the maximum possible, and at the same time it ensures that the universities are not inclined to question either the underlying philosophy or the likely ultimate outcome.

There are legitimate doubts about both.

First, it is stretching things a bit to assume there is much educational breadth to be obtained by spending three or four (and sometimes five or six) years pounding away at some esoteric research problem which, while it may indeed be interesting and even important, is of necessity highly specialized and is scarcely the stuff of general conversation.

Second, while it is probably true that most of the PhDs will find themselves a job on the basis of their paper qualification, this will be at the expense of those with fewer or lesser degrees. The over-abundance of PhDs has begun a chain of over-qualification which, quite apart from being wasteful in the economically rational sense, will ensure the existence of a fair number of people unhappy with their lot.

Third, there is emerging a large class of people who are struggling to pursue a research career via an endless treadmill of one-to-three year 'soft money' grants. Quite apart from the fact that the process in the long run is scarcely conducive to good research, the treadmill is not in fact endless. It probably finishes abruptly at the very awkward age of the early forties.

Fourth, one has to face the fact that the quality of the average PhD is not as good as it might be. There is no question that top-of-the-line research students are better than ever, and like the 'alpha-pluses' of old will turn out to be good scientists whatever are the short-comings of their supervision. But there is also no question that many research students are virtually carried through their degree simply because the system insists upon success. The failure of a PhD candidate, or a declaration of the unsuitability of a PhD candidate for research, is no longer a viable option for university departments.

In any event, one can present the background numbers about the oversupply of researchers to any group of scientists in virtually any discipline and they will agree with the overall conclusion but will invariably add a caveat. "Ah, yes" they will say, "but of course it isn't true in our particular field. Only last month for example we advertised for a research scientist to work in our group and there wasn't a single suitable applicant!" Superficially this state of affairs seems to violate some basic law of conservation. In fact it is a cause, a consequence and a fairly obvious example of the enormous changes in the practice and character of science over the last several decades.

Competition Gone Over-the-Top

The publish-or-perish syndrome has always been a fact of life for academics, and in earlier times was the subject of much amusement both within and without the fairly closed shop of the scientific world. These days it is no longer even slightly funny. It is a serious and deadly game whereby the promotion and survival of a research scientist is governed almost entirely by the number of his or her research publications. As a consequence, much of their work is no longer an exercise of enquiry in depth but is rather a contest to see who has best mastered the techniques of publishing the maximum number of research papers. Most of the techniques ensure that the work is confined to problems (or indeed to non-problems) which are easy to solve or where the answer is known beforehand. The career of any young scientist who cannot, by one means or another, publish at least two or three research papers a year is in quite serious trouble.

The dilution of content has become a survival technique for institutions as well as for individuals since the performance of entire research agencies is likely to be measured in terms of such things as the number of publications per dollar of funding. Both bureaucrat and scientist recognize the problem, abhor it in public, mutter darkly about the stupidity of the selection processes which succumb to it, and continue as before. To be fair, the bureaucrats every so often turn themselves inside out attempting to introduce the latest evaluation technique associated with the newest theory of management, but ultimately they are forced to return to the counting of papers because 'it is the only quantitative measure we have'. Researchers rather guiltily go along with the bureaucrats since they feel in their democratic bones that it is only fair that there should be some instantaneous and objective quantitative measure of their research productivity. Their penchant for fairness over-rides their common sense. Research output is one of those things which, at least in the short term, can be measured only by the subjective and essentially qualitative judgement of reasonably knowledgeable individuals. Democracy has nothing to do with it.

As something of an aside, the journal 'Nature' some years back reported an analysis of the publication rates of ten of the very top UK scientists who made a sizeable impact in their chosen field and who, towards the end of their career, became considerable movers and shakers in the scientific scene. In every case their rate of publication in the first ten years after their PhD was either very small or non-existent. It appears that they were able to sit back and work seriously for the length of time required to make a significant contribution. This in turn suggests what may be a basic rule of modern science – namely, that one can do significant research or can publish lots of papers. It is unlikely that one can do both at the same time.

In any event, no institution today can afford to appoint a research scientist who is trained in another discipline and would therefore need a year or three to become productive (that is, to produce research papers) in the new field. New appointees must have direct experience in the precise field of the institution. Here then is the basic reason for the puzzling universality of the caveat concerning the over-supply of PhDs. Finding an exact fit between newly graduated PhD and actual job requirement is indeed very difficult. It is also exactly what is required to ensure narrow-mindedness of scientific outlook, mediocrity in research, and a scientific literature so vast and overladen with the minutiae of the unimportant that most of it is never read by anyone. Which is quite a change from the middle of last century, when it was drilled into the aspiring scientist that the very worst thing to do was to "spend one's life repeating one's PhD". Change of field, and indeed complete change of discipline, was regarded then as one of the essential mechanisms for maintaining the quality of scientific research.

The final outcome of all this is that modern scientific research is incredibly competitive, which of itself is a fair indication of the overproduction of practitioners. And while there is nothing inherently wrong with a certain amount of competition, there is certainly a problem when it begins to affect very badly the quality of the ultimate output. Among other things it is preventing scientists and scientific institutions from making the mistakes which are an essential part of worthwhile research.

Concerning the Character of Science

Snow's thesis that scientists and literary intellectuals are unable to talk to each other was not all that original, but it was he who brought the problem into the open. He recognized, and Sir Peter Medawar laid it out rather more specifically in later years, that science is inherently incapable of answering many of the more fundamental questions about the world. These are the "Where have we come from? Where are we going?" and, in particular, the "Why are we here?" sort of question. They are the province of literature and philosophy rather than of science. And while it is a little sad from the purely academic point of view that this difference between the disciplines detracts from efforts to sort out the intellectual order of things in general, Snow's main point was that the existence of a fence between the understandings of one side by the other is actually a dangerous state of affairs. It is after all the literary philosophers who have to put scientific advances into the context of society and its times.

The man in the street can normally read the original sources of literature and philosophy. They may be heavy going, but he can almost certainly understand them provided he stays awake long enough. Scientific literature is not like that. It can be highly esoteric and anyway is probably written in its own language, so that useful communication to the general public requires an interpreter who can straddle both sides of the fence. The interpreter may of course be the researcher himself, and it is often the case that the most brilliant scientist is also a great communicator. On the other hand the average scientist (and he is most of them) has to be very careful about 'playing down' to the public because his colleagues can be quick to label him a lightweight. And anyway he is probably not very good at talking to ordinary people about his work. The interpreting role can be performed most easily, and more safely, by those who don't have a research reputation to lose.

The need for professional interpreters has now become entangled with the need for public relations consultants who can find their way around the competitive world of scientific politics. It is not too surprising that a new breed of interpreter has evolved who does a lot more than simply translate science for the masses or for Snow's first culture. He has to be broad in his knowledge so that he can interact with lots of disciplines, and does not always have much of a feel for the depth, detail or scientific significance of the research he is paid to report. In common with consultants the world over, he has to persuade his client (in this case the scientist at the bench) to do most of the actual writing and explaining for him. He simply re-packages it all in fancy wrapping and goes about a process of selling the stuff both to the public and to the inhabitants of the corridors of power.

Now there is nothing immediately sinful about the process except that it probably wastes a lot of research time. On the other hand it needs to be recognized that the interpreter/consultant is selling the wrapping rather than the product, and furthermore is selling it into a nearly saturated market. The media are awash with press releases from the scientific institutions, and it is really quite extraordinary how many breakthroughs seem to be occurring which, of course given money and a further five years research, will undoubtedly prove of enormous benefit to society. There are lots of TV and radio shows about science, many of them designed to attract young people and most of them deriving ultimately from the public relations needs of an increasingly competitive activity. Very few of them actually advance the public's knowledge of science itself, but rather of the political context in which it exists. In such a context, the concepts of truth and proof can become a little bent, and bent in such a way that even scientists find it difficult to recognize what is happening. The literary people can easily be led astray.

This brings us back to the fence between the cultures. Great literature is great mainly because it manages to sell a point of view about an issue for which there is in fact no straight yes or no answer. It involves advocacy. It is biased in the sense that one expects a lawyer to be biased when he is putting forward his case in court. Science on the other hand normally sticks to the answerable questions, and can afford the luxury of insisting on truth and proof in its reporting. It is supposed to provide impartial answers.

There have been many attempts throughout recent history to draw the two cultures together. The material success of science ensured that the attempts were almost always in the direction of making literary philosophy more scientific rather than the other way about. And indeed there has been a change, although not perhaps in the first culture. The disciplines which Snow suggested might be a third culture have developed into what we now call the soft sciences. They include economics, history, government, archaeology, psychiatry and so on, all of which (with the notable exception of psychiatry according to Medawar who must have been bitten by one of its practitioners) have benefited considerably from reference to the techniques and philosophy of science. Fortunately perhaps, the move to change the character of literature, philosophy and practical politics has succeeded not at all. They would not amount to much if their fundamental right to advocacy and bias was removed.

Certain areas of science are beginning to adopt some of the first-culture characteristics. It is not anything particularly deliberate, but is rather a consequence of a necessary awareness of the politics which surrounds the modern competitive nature of science. It is particularly obvious in the environmental sciences since they deal more often than do other disciplines with issues that have a philosophic background related to the 'why are we here' sort of question, and with the best will in the world an element of bias can enter into the work. It is not such a terrible thing provided the bias is confined to the selection of topics on which to spend the research effort. It becomes extremely worrying when it gets to the stage, as it has for instance in the climate change debate, that 'scientific consensus' is used as an active tool in the shaping of public opinion. Inevitably the tale that is publicly told draws down upon the hard-won capital of the scientific reputation for impartial judgement. Inevitably the researchers themselves are forced into the first-culture mode of advocacy for the consensus opinion.

And inevitably also, once such a consensus is established it becomes extremely difficult for a scientist to do what he or she is paid to do – namely, to be sceptical about accepted wisdom. The large-scale problems addressed by modern science usually involve extensive and expensive collaborative research effort. Checking or repeating the results requires a similar effort. What granting body looks favourably on a proposal designed effectively to repeat already published research supporting a formally accepted consensus? The problem gets back to competition and the publish-or-perish syndrome. Scientific papers are supposed to record new and original findings, and very few agencies will fund programs which, on the bureaucratic face of it, run the risk of establishing nothing new. More to the point, the standard of proof required for acceptance by a reputable journal of a paper which goes against a formally established consensus can be raised, albeit sub-consciously, very high indeed.

Many problems are so diverse that no one scientist can fully comprehend every thing about the subject. He has to retreat into detail. He has to define for himself a small element of research which, with a certain amount of spin from the scientific interpreters, can be fitted into the wrapping which has been evolved to suit both public opinion and the source of his money. More important, his belief in the validity of public announcements by his masters about the overall issue is determined entirely by his trust in the integrity and skill of the scientists working on other aspects of the problem. His own opinion on the overall issue is not an independent piece of information, and the concept of scientific consensus (namely, that most scientists agree with the public announcements so those announcements are probably true) can be something of a nonsense. The consensus is really about the trustworthiness of other scientists rather than the problem itself.

Almost by definition, so-called 'public good' research is not a great money spinner in a competitive world, and new criteria for success have evolved. One criterion nowadays built into the performance measures of public-good and environmental research laboratories is that the work must make a political impact – that it must be shown to be a significant factor in determining the outcome of political decisions. (How this can be done in any practical situation is a secret known only to the proponents of the new managerialism, but there we are). The criterion is an open invitation for scientists in these fields to think and behave as paid-up members of Snow's first culture – or at least to behave as paid-up members of its operational arm of everyday politics.

And some scientists and their interpreters – not many, but certainly rather more than enough – have taken to the process as ducks take to water. They have become experts for instance at generating doomsday scenarios which capture the public attention. Their international connections and their access to reputable international organizations give them a political clout which the typical activist in other spheres can only dream about. At the back of it all is the fact that many of the public-good and environmental sciences are a sort of engineering mix of bits and pieces from the more established physical and biological sciences, and such mixes are not really likely to produce advances in our understanding of the fundamental ways of the universe. Political involvement can be attractive because it provides a different, and potentially much easier, avenue to fame and fortune.

The Other Side of the Coin

It is all very well to bemoan the political overtones associated with consensus science, but the fact remains that governments have to make serious decisions based on scientific input. How else can they operate other than to go to the scientific community and ask "What is your best estimate about such and such?". And while there is a sizeable and increasing number of scientists who rather enjoy the political process, the further fact remains that the vast majority find it extremely difficult to make any statement at all about questions which do not at this time (and maybe not at any time) have a definitive answer. They are frightened, and perhaps rightly frightened, that dilution of the reputation of science with first-culture responsibility is too high a price to pay. On the other hand, governments are annoyed, and perhaps rightly annoyed, when the scientific community refuses to give a best-guess answer to a question needing a quick response, but mumbles instead about having to do more research on the matter, and please, can it have more money for the purpose.

So one can perhaps argue that it is really quite a significant advance for the scientific community even to begin thinking of providing such things as consensus views on difficult problems. Certainly, something of the sort will become more and more common as research is forced into larger and larger issues with more and more public involvement. The trick will be to ensure a clear understanding that the pushing of barrows is strictly a first-culture activity best left to first-culture practitioners. It isn't necessarily a good thing for the fence between the cultures to be lowered as far as possible, since the more public and political is the activity, the more obvious to the bureaucrats becomes the need for control. Mickey mouse mixtures virtually demand mickey mouse systems of management.

CHAPTER 6

WHY IS IT SO?

"The next great task of science is to create a religion for mankind."

Lord John Morley of Blackburn (1838-1923)

Most of us loudly abhor the stupidity of political correctness when we fall over it, but usually try to keep within its bounds when making our own public conversation. This may not say much for our courage, but survival in an overly sensitive society seems to require something of the sort. Perhaps the main difficulty is to find a universally accepted definition of a politically correct statement. Most of us would be hard put to it to say precisely what we ourselves mean by the term, let alone what is meant by anyone else, so we don't really know whether to laugh or to cry at any particular example of the disease.

A Civitas monograph a few years back by Alexander Browne is worth reading. He provides a definition that typecasts most politically correct arguments, and allows us to recognize new ones for what they are. He shows us that there is usually far more reason to cry about them than to laugh. He shows us in fact that political correctness can be highly dangerous.

His definition is based on the idea that many people prefer to think in black and white rather than in shades of grey. In particular such people think of humans and their social organizations as either good or bad. There is no middle ground. And in order to simplify the thinking process about any specific issue (or indeed to eliminate the need for thinking about it at all) they have come to the convenient conclusion that the ideas and actions of the 'have-nots' of any situation are good by definition, and those of the 'haves' are bad.

It is scarcely an original conclusion since it has roots in almost all of the religious and social philosophies of the world. What makes it stand out a little from the crowd is its absoluteness – its complete dismissal of the possibility that there can be exceptional circumstances when at least some thoughts and actions of the 'haves' might be good and at least some thoughts and actions of the 'have-nots' might be bad. This form of absoluteness is the defining characteristic of PC argument in that the side of the 'have-nots' becomes automatically apparent as the side of the angels. Publicly supporting that side – the politically correct side – gives an immediate feeling of virtue. And feelings of virtue, it seems, have become a valuable 'good' in this economically fortunate era when people have leisure to indulge themselves. Acquisition of such a 'good' in such a manner is highly rewarding and requires nothing at all by way of mental exercise.

The bottom line is that the addictive nature of publicly expressed virtue ensures that many people prefer to accept arguments based on political correctness rather than on common sense or on scientific observation. Should that preference become universal, society in the future will find it much more difficult to define its significant problems. It will tend – as is already happening in the context of global warming – to waste vast quantities of its human and physical resource on strange forms of social engineering that have little to do with real issues and nothing at all to do with the enjoyment of life.

The hugely popular doomsday cult of climate change and global warming has developed virtually all the nasty aspects of political correctness, of which the most obvious is an unthinking, and highly public, intolerance of the opposing view. And it is certainly true that we are on the brink of wasting in its name mind-numbingly large amounts of money and resource on strange unproven exercises of social engineering.

The Hidden Agendas

One might take more notice of the greenhouse global warming scare campaign if it were not so obvious that its most vociferous supporters have hidden agendas. There are those who are concerned about preservation of the world's resources of coal and oil for the benefit of future generations. There are those who, like President Chirac of France, look with favour on the possibility of an international decarbonization regime because it would be the first step towards global government. There are those who, like the socialists before them, see international action as a means to force a re-distribution of wealth both within and between the individual nations. There are those who, like the powerbrokers of the European Union, look upon such action as a basis for legitimacy. There are those who, like bureaucrats the world over, regard the whole business mainly as a path to the sort of power which, until now, has been wielded only by the major religions. More generally, there are those who, like the politically correct everywhere, are driven by a need for public expression of their own virtue.

Of course there is nothing wrong, or at least not much that is wrong, with the ideals behind any of the above agendas except perhaps the last couple on the list. But the battles over them should be fought on their own merits rather than on the basis of a global warming crusade whose legitimacy is founded on still-doubtful science and on massive slabs of politically correct propaganda.

It is generally assumed that climate scientists themselves are not pushing the global warming barrow simply because of their interest in some other agenda.

Well perhaps. They have been so successful with their message of greenhouse doom that, should one of them prove tomorrow that it is nonsense, the discovery would have to be suppressed for the sake of the overall reputation of science. In a way, the situation for them is very similar to that of the software engineers who sold the concept of the Y2K bug a decade ago. The 'reputation stakes' have become so high that it is absolutely necessary for some form of international action (any action, whether sensible or not) to be forced upon mankind. Then, should disaster not in fact befall, the avoidance of doom can be attributed to that action rather than to the probability that the prospects for disaster were massively oversold.

It is more-or-less standard procedure to label as biased those scientists who are sceptical of the dangers of global warming because of their status as knowing stooges of the energy industry. This is at least an admission of the belief by global warming advocates that scientists as a class are in fact quite capable of having their judgement corrupted by political and monetary interest. That being so, the mistrust of sceptics is a peculiarly lopsided assessment. The potential for bias towards the politically correct, and massively government supported, side of the greenhouse debate is the greater by orders of magnitude.

Where in Australia for instance can be found the industry equivalent of the ten million dollars made available to the Climate Institute a few years ago specifically for spreading the word about global warming disaster? To say nothing of the nearly ninety million dollar annual cost of running the federal government's Department of Climate Change. Where in the US can be found the industry equivalent of the massive international propaganda campaign mounted by Al Gore as a follow-up to his film 'An Inconvenient Truth'. Pity the politicians who, we presume, are trying their best to make an informed decision on the matter. Of course politicians realise that those clamouring for their attention on any particular issue usually have other un-stated agendas. But they may not recognize that scientists too are human and are as subject as the rest of us to the seductions of well-funded campaigns.

Uncertainty and the High Moral Ground

Perhaps the major characteristic of the science of global warming is the uncertainty associated with just about every aspect of it. Setting aside the difficulties of models and their forecasts that we discussed in earlier chapters, the actual measurements of trends in the environment are invariably plagued with inaccuracy. More to the point, in climate research there is no such thing as a controlled experiment – except of course in the overall sense that the earth-atmosphere system is already being subjected to a forced change of atmospheric carbon dioxide, and ultimately no doubt we will see what the response of the system turns out to be. Even then it may be that we will not necessarily be able to ascribe the response to the specific cause of CO_2 forcing. A researcher in climate is always faced with the problem, perhaps more familiar to biologists than to physicists, of disentangling a multitude of causes for any particular effect. The study of climate is a statistician's paradise.

Now statistics is a highly developed discipline that is at its best – or perhaps one should say at its most understandable – when it is used in the development and interpretation of controlled experiments. It is an essential tool for the interpretation of uncontrolled experiments as well, but as the degree of control gets less and less, the required statistical analyses become more and more esoteric. They become more of an art than a science.

The statement is in no way intended as a denigration of statistics or statisticians. Rather it indicates that in such circumstances the identification of real relations between cause and effect requires not only a great depth of knowledge and an intuition about what is going on, but also a level of scepticism that is difficult to maintain from within one's own research field. There has to be an ability to divorce oneself from unconscious preference for the desired result. There has to be an ability to recognize and admit when the available data may not be enough to give a definitive answer one way or the other.

All of which boils down to a situation where the general acceptance of a scientific proposition – even among the elite of the scientific fraternity – can become largely a matter of which side of the argument provides answers most people would want to hear. And this in turn, particularly in a research field that has the potential impact on human welfare of global warming, can be determined by agendas and beliefs that have nothing directly to do with the science of the matter.

This brings us back to the story we told in the Introduction about the reaction of scientists to scepticism about the 'hockey-stick' proxy record of global temperatures of the past. It seems these days that one is not allowed to be openly critical of a published result unless one is an expert in the particular narrow field of science. The fact that such a philosophy is directly at variance with scientific opinion of an earlier era is completely forgotten. It is forgotten that many of the major advances of the past were made by scientists who were expert in disciplines unrelated to that of the original problem. It is forgotten that, in days of yore, a young PhD graduate was encouraged – and occasionally forced – to move into a research field as different as possible to that in which he was trained. And indeed, a classic example of the advantage of new blood was the appearance out of the blue of a geological statistician who managed to sort out the 'hockey stick' debacle.

It also seems these days that the only proper course for a sceptical general scientist is to write up his objections in a manuscript that can be subjected to formal 'peer review' before publication in an appropriate research journal.

Good in principle maybe, but in practice not always a course of action that is particularly rewarding.

Once a point of view has entered the literature, the barrier to publication of a contrary opinion becomes very high. Inevitably a sceptical argument will have much the same degree of doubt and uncertainty as the original proposition. The existence of uncertainty can then be used, either consciously or subconsciously, to support the status quo. The sceptic finds himself in the position of having to prove the original proposition wrong before a journal will accept his work for publication. The particular difficulty in climate research is that proof of anything at all is in fairly short supply.

We mentioned in the previous chapter that the promotion of a research scientist these days is even more dependent on the number of his or her published research articles than it was in the past. Vast numbers of journals have appeared out of nowhere to cater for the requirement. As a consequence, the powers-that-be within the science community have instituted both formal and informal pecking orders that define the worth of any particular journal. The object of the game as far as authors are concerned is to get their papers in the most prestigious of the journals, since ranking to some extent now counts along with number in the promotion stakes. It is assumed that the higher the ranking of a journal then the higher the quality of the papers that appear in it.

The assumption is not entirely wrong. A good journal has a vested interest in keeping its reputation for quality, and tries to ensure that its reviewers – that is, the scientists within the particular field of research who are asked to read and comment on the worth of a manuscript before it is accepted for publication – are fairly tough in their assessment. On the other hand, and particularly in the prestigious journals, one has to bear in mind that the reviewers will usually be part of the research elite in that particular field and will probably be sympathetic with the existing viewpoint. Indeed they will probably be the ones who established the viewpoint in the first place.

Modern climate science is so complex that it is normally quite impossible for a reviewer to check the actual results described in a particular research manuscript. The results may be from the output of an enormous computer model which has been developed by teams of specialists over the years, and the reviewer has no way of running that (or indeed any other) model to perform cross checks on the conclusions reported in the manuscript. The results may be from sets of measured data gathered by others over the years which, for one reason or another, will have been manipulated in various ways to remove what is assumed to be erroneous information. There is little chance of a reviewer having the time or the inclination to examine the data sets, check the results, or figure out whether the manipulations (even if they could be identified) have really introduced more certainty into the final conclusions.

A reviewer is normally not paid for his work. With the best will in the world, he is able to spend no more than a few hours examining any particular manuscript. He is able to do little more than see that the story being told is superficially coherent and makes no obvious errors of fact. He will check whether the paper formally refers to other, already published, papers that are relevant to the topic. Almost inevitably he is forced into making an assessment based only on whether or not the new results are supported by those reported in already published journal articles.

In short, there exists a process of natural selection tending to ensure that research publications conform to the existing norm. It has come about because reviewers of research manuscripts generally have no practical means of assessing new results in terms of a very fundamental criterion – namely, that the results are reproducible by others. Reproducibility is particularly important in a field of science where there is not a lot of opportunity for the checking of theory by experiment.

Indeed, the issue of reproducibility looms very large as the main factor leading to tremendous distrust of the motives of the global warming scientific establishment by those on the periphery. A good example of the problem concerns again the outcome of the famous 'hockey stick' debate already mentioned several times in earlier chapters.

There has existed for some years now a weblog called 'climateaudit.org' which is run by Stephen McIntyre. Mr McIntyre was one of the two Canadians who did the uncommon thing of actually trying to reproduce the hockey stick graph of global temperature that was originally reported in the journal 'Nature' back in 1998, and which became a centrepiece of the third IPCC report. They couldn't reproduce the results, and went on to show exactly what was wrong with the methodology used to derive them. Basically it seems that the statistical technique used by the hockey stick authors was such as to 'mine' data sets of tree-ring thicknesses for those particular sets that led to a hockey-stick shape in the graphs of past temperature. This was presumably not a case of deliberately choosing a technique so as to get a desired result, but as mentioned earlier, statistics can be tricky. There were other issues as well, among them an extreme dependence of the results on that portion of tree-ring thickness data from around the world that was derived from bristlecone pines in North America. The ring thicknesses of bristlecone pines are known to be highly questionable as a proxy for temperature.

Mr McIntyre and his colleague Ross McKitrick had enormous difficulty getting these matters into print in the formal research journals. They succeeded eventually, but the whole business was a classic demonstration of how the scientific community tends to close ranks in support of the attitudes and published results of its currently favoured sons. It was largely as a consequence of this demonstration that Mr McIntyre began his weblog.

Through it, he has regularly analysed and commented in detail on a whole series of research papers from the hockey-stick research establishment of dendrochronologists. It has led among other things to formal enquiries into the hockey-stick matter by the US Academy of Sciences, and even by a committee of the US House of Representatives. To the obvious discomfiture of his targets, Mr McIntyre turned out to have the time, energy and personal resources (he is retired from a career in the mineral exploration business) to pursue the matter of reproducibility of results to an extent quite foreign to various elements of the climate-science community. Not to put too fine a point on it, he frightened the hell out of them.

He had extraordinary difficulty in abstracting from the hockey-stick authors either the original data or a detailed description of the analysis that led to the hockey stick results. These are the sorts of things which, in principle, the research journals require to be archived in publicly-accessible computer files if they are not already in the journal article itself – this precisely so that others can test the reproducibility of published results. Over the years since then, it turns out that it is very rare indeed for such files to be made available by authors from the climate-change fraternity. Or at least it is rare for them to be made available in a form that is sufficiently understandable to be used. Suffice it to say that Mr McIntyre has of necessity spent a goodly fraction of his time attempting to get at original data via formal freedom of information requests to climate research agencies. Not the most popular thing to do, and not generally very successful. It turns out as well that it is extraordinarily difficult to get the journals themselves to insist that their authors abide by their archiving policy. It seems that both research agencies and journals are very good at looking after their own.

The climate research establishment spends a good bit of its time patching up its mistakes as they are exposed one after the other by Mr McIntyre. At the same time it plays a strange game of ignoring his existence. Which is sad in a way, since by now he is probably *the* world expert in the field and should in principle be involved in helping to abstract the best information possible from the various sources of proxy data concerning temperatures over the last 1000 years. The official attitude towards him is summarized by the comment of one researcher who, when asked for the basic data that were used in some particular piece of research, came out with the following gem. "Why" he said in an e-mailed reply "should we make our data available to you, when your objective is to find something wrong with it?" Surely one of the classic examples of a remark by a scientist who simply cannot hear what he is saying.

Mr McIntyre has expanded the interests of his weblog to cover quite a number of the more contentious issues of climate science. The research establishment hates it, and rarely contributes directly to the scientific arguments carried on there. The reason is obvious enough. The weblog has an enormous audience – for the last three years it has had over 6 million 'hits' per year – and is developing into an effective counter to the various biases and obstructive behaviours which now surround the regular scientific literature.

The working scientist knows all too well how the chips are stacked in the unending fight to maximize the number of his papers that can be cited with his next application for promotion. He knows that the overwhelming balance of reviewers will be on the side of the angels – in the present context on the side of the main conclusions of the IPCC – so will be very reluctant to draw conclusions which are at variance with the common view. Safety indeed lies in numbers.

And the more idealistic sceptical scientist also can do his sums. He knows that the chances of publishing his views in a high-quality journal are very slim. He is more-or-less forced into the arms of non-refereed journals, or at least into the arms of journals way down the pecking order of official ranking. His results can then be dismissed by the cognoscenti as sub-standard and of no account.

One of the powerful forces bearing on climate scientists is a sort of misplaced loyalty. It may not be directed specifically towards their employer, although it has amounted to that ever since the major research establishments began to follow nicely along politically correct lines. Rather, it appears as support for the general reputation of science. It appears as support for the particular view of climate change that is being sold by scientific activists to the heads of their organizations and to the public at large.

It appears because most scientists simply cannot believe that their colleagues would deliberately oversell a scientific conclusion for the benefit of a political cause. Dishonesty of that nature would fly in the face of everything that the rather idealistic typical scientist has been taught about his profession. It follows from this belief that the activism of certain climate scientists cannot really be activism, but instead must be a dispassionate statement of truth by members of the profession who happen to know more than anybody else. The integrity of science itself demands that their view of affairs must be openly defended.

And so we hear from our colleagues: "How dare you put your sceptical view to that external advisory committee. It is essential that we present a unified voice on the need for action on climate change. Governments cannot make the correct decisions if their agencies allow confusing messages to get through the system and into the ears of politicians. The issue is too important to allow dissent."

Or words to that effect. Sir Humphrey Appleby would be proud of them. Once again it is staggering that so many of the climate science fraternity cannot hear what they are saying. It is staggering also to hear the loud declamations by heads of research agencies about the absolute freedom of their research staff to make public comment within the areas of their expertise. Presumably the agency heads believe what they are saying. If so, they quite fail to understand the forces governing the attitude of their staff. Or to be less than charitable, perhaps they understand them extremely well and prefer to keep quiet about the matter.

Suffice it to say there is much to the suggestion by the scientifically literate author Michael Crichton in one of his lectures in the US a few years back:

> Sooner or later, we must form an independent research institute in this country. It must be funded by industry, by government, and by private philanthropy, both individuals and trusts. The money must be pooled, so that investigators do not know who is paying them. The institute must fund more than one team to do research in a particular area, and the verification of results will be a

foregone requirement: teams will know their results will be checked by other groups. In many cases, those who decide how to gather the data will not gather it, and those who gather the data will not analyse it. If we were to address the land temperature records with such rigor, we would be well on our way to an understanding of exactly how much faith we can place in global warming, and therefore with what seriousness we must address this.

CHAPTER 7

CONCLUSION

"In the school of political projectors, I was but ill entertained, the professors appearing, in my judgment, wholly out of their senses; ...and confirmed in me the old observation, that there is nothing so extravagant and irrational which some philosophers have not maintained for truth."

Jonathan Swift in his 'Gulliver's Travels'

In one limited sense the members of the "do something about global warming" lobby are correct. If humans insist on giving the atmosphere an extra dose of carbon dioxide, then indeed one can expect Earth's surface temperature to rise. To be strictly accurate, we should say that its temperature will be higher than it would have been otherwise. Either way, it doesn't take a lot of physical knowledge and insight to accept the statement. It is rather the equivalent of saying that if one hits something with a bat then that something will respond. So it is true, as the lobby delights in telling us at every opportunity, that there is no longer much argument among scientists about the existence of the greenhouse global warming phenomenon. There never was.

The consensus goes no further down the chain of political correctness than this. It is rather naughty of the greenhouse lobby either to say outright, or to imply by judicious omission, that it does.

It has not been solidly established, and it is certainly not accepted by the majority of scientists as proven fact, that global warming from increased atmospheric carbon dioxide will be large enough to be seriously noticeable - let alone large enough to be disastrous.

Imagine the response of a well-bedded concrete post when belted by a relatively small bat. In a situation where the post has been around a long time and has in the past survived the beatings of lots of much bigger bats, the chances are that it won't move much.

More than thirty years of well-funded international research directed specifically at the climate-change problem have brought us no nearer to an estimate of future temperature rise than to say, rather feebly when one thinks about it, that the global-average increase over the next century may be somewhere between one and several degrees Celsius. Thus say the various computer models, whose simulations even of present climate fall into the 'reasonable' range only by dint of forced tuning of many of the pieces of input information. There are no means of experimentally checking the overall predictions of future climate change – basically because our knowledge of past climate is not precise enough. Furthermore, it should be remembered that the "one to several degrees" range covers only a limited set of the results obtained from all possible variants of climate model. The choice of that particular set derives from what might be called seat-of-the-pants statistics – the sort of statistics practiced by members of a committee dedicated to producing figures which on the one hand are interesting and on the other are not so over-the-top as to be rejected by their peers. Suffice it to say that there are more than enough pitfalls associated with the application of statistics to actual measurement. The pitfalls are multiplied enormously when applied to various manifestations of pure theory.

Even accepting for the sake of argument that some significant degree of global warming may be observed in the future, it is certainly not the consensus of the majority of scientists that the actual impact on humans will be significant – or indeed that it will be detrimental. The bottom line here is that computer models have no provable skill at forecasting the change of regional and local climate even if we accept that they may say something sensible about global averages. In particular it may be that things like the continental, regional and local averages of rainfall are *inherently* unpredictable. Therefore the models are in no position to tell us anything of the impact of climate change on any particular aspect of human endeavour. Instead one must resort to all sorts of 'what if' scenarios, virtually all of which have no justification other than that they are easy enough to sell as doomsday forecasts to politicians and to the public. "Where it is dry we will get more droughts. Where it is wet we will get more floods. Where there is disease, it will spread. Where there are people the sky will fall in." Such predictions are tailor-made for the mournful tones of the politically correct reformers of mankind. They are now accepted without a murmur of dissent by a large fraction of western society.

The trouble is that the uncertainty inevitably associated with the chaotic behaviour of climate works both ways. It may be impossible even in principle to substantiate a doomsday forecast, but it is also impossible to prove anything to the contrary. So the winning side of any argument about the matter will inevitably be the side with the loudest collective voice. In any event, should the doomsday scenario indeed fail to inspire fear and trepidation because it cannot be substantiated, one can always fall back on its unspoken basis – namely that "all change is bad".

And so we come to some final comment on the questions raised or implied in the introduction. Why is it that the scientific community has become so one-eyed in its public support for the disaster theory of climate change? Why is the scientific community taking such an enormous risk with its reputation?

In fact, the short-term risk to the profession is probably not all that great. In view of all the uncertainty inevitably associated with argument on either side of the fence, it is not likely that anyone will be able in the near future to prove absolutely that any particular forecast of climate change is nonsense. It has taken the IPCC more than 20 years to develop a story which, though replete with uncertainty at just about every level, is coherent enough to be sold to the public at large. Perhaps more to the point, the story is complex enough to be virtually unarguable by anyone or anything other than a fully-fledged research institution specifically assigned to make that argument. Thus it is unlikely – not impossible, but unlikely – that an individual somewhere will produce a single scientific result powerful enough to blow the idea of disastrous global warming out of the water. It is even less likely that a national government would risk the anger of its scientific establishment by creating a research institution – it would have to be a very large research institution – designed solely to perform a large-scale critical audit of the scientific bases of the forecasts of climatic doom. While the suggestion along these lines by Michael Crichton quoted in the previous chapter was sensible enough, one has to suspect he didn't really hold out much hope that such an institution would ever come to pass.

On the face of it, the long-term risk to the profession is much greater. In fifty or a hundred years the forecasts of doom will have been tested and, with any luck, proved wrong. But by then the leading role of the scientific community in upsetting the global economic system will probably have been forgotten. The scientist of that time will be able to dig into the archives and find various quotes to the effect that "on such and such an occasion, this or that scientist spoke publicly about the uncertainty of the climate forecasts". He will therefore be able to maintain with his hand on his heart that it was not the fault of scientists that society went overboard on the matter. Rather, it will have been the fault of the environmentalists and politicians who misinterpreted the scientific results for their own nefarious purposes. Loud enough repetition of statements along these lines should effectively obscure the existence in the past (that is, in the 'now') of a carefully calculated campaign to trade scientific reputation for political action.

By then as well, there will be enough 'wiggle room' to evade serious enquiry as to why scientists rarely bothered to refute in public the more fantastic of the scenarios for climatic disaster. "It was not our job to protect the public from misinformation" they will say in the year 2109. Die-hard global warming scientists make that comment even today. Strangely enough, they are not nearly so coy when it comes to refuting ideas to the effect that things might not be as bad as they are painted.

As to the 'why' of the business, there are a fair number of very strong forces at work to encourage the interpreters of climate science to overstate their case. To a large extent the forces are at work also on the scientists themselves. As with all religions, woe betide those demented souls, scientists or not, who are so deluded as to question the beliefs of the politically correct.

It is worth remembering that among the interpreters are the scientific administrators – in particular the managers of research institutions who by virtue of their office are the official spokesmen for the views of their organizations. Their words carry tremendous authority with the public because it is assumed they have a deep understanding of the science for which they are responsible. Sadly, in the modern era of management, that assumption can be way off the mark. They may have little real knowledge of science, and are as subject to the necessities of political correctness as the rest of us. Indeed, perhaps rather more than the rest of us. Many of them have been appointed to their position precisely because of their 'feel' for the views and needs of the community rather than their 'feel' for science.

A number of pragmatic reasons for a sub-conscious bias by the ordinary bench scientist towards the politically correct have already been mentioned. Basically they boil down to the need to eat. Fame and fortune in the research profession depend largely on artificial measures of success related to the quantity rather than the quality of research publications and of grants. Undoubtedly the system rewards conformity to the popular view when outcomes are determined by consensus rather than proof.

There are also a number of less pragmatic reasons for bias. Among them are the other agendas mentioned in the previous chapter. But perhaps the saddest and most deeply hidden is related to the fact that much modern research can be intensely debilitating to the scientist concerned. The reward system of his profession forces him to spend a great deal of his time researching safe topics whose importance in the grand scheme of things is virtually nil. He can be reasonably certain that work of this type, when published, will probably never be read by anyone. Persuading himself that it is nevertheless significant and worth doing requires a tortuous and painful exercise of self-delusion. In such circumstances it is an immense relief to be associated with an international programme which, whatever one might think of its aims and politics, at least has high and popular moral purpose. It restores his pride.

The above rather rambling discussion about bias in the global warming story has basically circled round a point made earlier – namely, that the potential for bias is overwhelmingly toward the politically correct. If for no other reason, the money lies on that side of the fence. Perhaps the most interesting, and probably unanswerable, remaining question about it all is how a belief in climatic doom became politically correct in the first place. Conspiracy theorists would probably favour the idea that it was all planned 30 years ago by some small, shadowy, secret organization bent on destruction of the world's social order. Personally I would rather believe that, given the human addiction to tales of collective guilt, there is no need to invoke conspiracy as part of the explanation. The path to the final outcome was inevitable from the start.

Finally it is worth re-making the point that a situation has emerged wherein the politicians who must make decisions on the matter of climate change are being deprived of a basic tool of their profession – namely, access to a diversity of advice from the 'scientific technological elite'. President Eisenhower was indeed extraordinarily prescient on the matter, and it is worth providing the full context of the quote used as a lead-in to Chapter 1. Remember that this was a speech made way back in 1961:

> Today, the solitary inventor, tinkering in his shop, has been overshadowed by task forces of scientists in laboratories and testing fields. In the same fashion, the free university, historically the fountainhead of free ideas and scientific discovery, has experienced a revolution in the conduct of research. Partly because of the huge costs involved,

a government contract becomes virtually a substitute for intellectual curiosity. For every old blackboard there are now hundreds of new electronic computers.

The prospect of domination of the nation's scholars by Federal employment, project allocations, and the power of money is ever present – and is gravely to be regarded.

Yet, in holding scientific research and discovery in respect, as we should, we must also be alert to the equal and opposite danger that public policy could itself become the captive of a scientific-technological elite.